70 0173392 9

British Institutes' Joint Energy Policy Programme
Policy Studies Institute
Royal Institute of International Affairs

GAS'S CONTRIBUTION TO UK SELF-SUFFICIENCY

Jonathan P. Stern

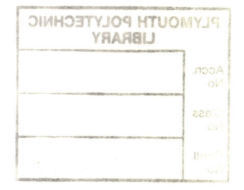

 Heinemann Educational Books

Heinemann Educational Books Ltd,
22 Bedford Square, London WC1B 3HH
LONDON EDINBURGH MELBOURNE AUCKLAND
HONG KONG SINGAPORE KUALA LUMPUR NEW DELHI
IBADAN NAIROBI JOHANNESBURG EXETER (NH)
KINGSTON PORT OF SPAIN

First published 1984

ISBN 0 435 84343 5

Printed in Great Britain by
Biddles Ltd, Guildford, Surrey

CONTENTS

Foreword
Energy Self-Sufficiency for the UK?

This paper is one of a series of reports being published in 1984 as a result of a research project by the British Institutes' Joint Energy Policy Programme on the question of how far the maintenance of self-sufficiency in energy is a desirable or practicable goal of policy for Britain. Each of the main energy sources - coal, oil, gas, and electricity (including nuclear energy) - is being examined in turn, in addition to conservation as an additional energy resource. These reports seek to describe the present contribution of each fuel to self-sufficiency, and the potential for extending that contribution into the future, together with the costs and benefits involved.

Two further volumes are planned, drawing on the basic information contained in the first five. One will seek to consider the interaction between decisions on the various fuels and to draw together the policy issues. The other will examine 'The Economics of Energy Self-Sufficiency' from a theoretical aspect.

In each of the first five papers the authors have sought to distil the welter of information available, including such sources as the Sizewell Inquiry into the construction of a nuclear power station, and the Monopolies and Mergers Commission report on the coal industry, and also to fill some of the remaining gaps in public knowledge of energy matters and to point to the policy issues that arise. In each case, they have had the benefit of advice from Study Groups of individuals expert in their particular fields and our thanks are due to those who took part. However, the views and interpretations presented are those of the authors, and not necessarily those of the members either of the study groups or of the institutions that sponsor the British Institutes' Joint Energy Policy Programme.

Comments on the contents of these papers, or suggestions as to how the policy issues should be tackled by government, industry or others, will be welcome. Every effort will be made to take account of such comments in preparing the final overall volumes.

We hope that this series will make a contribution to the development of policy by government and industry, and to public debate on energy issues.

Robert Belgrave.

Dry, non-associated natural gas: natural gas originating from structures where only gas is produced.*

Associated natural gas: natural gas originating from underground structures producing both liquid and gaseous hydrocarbons. The gas may be dissolved in crude oil (solution gas), or in contact with gas-saturated crude oil (gas cap gas). In such structures, gas production rates will depend on oil output, with oil usually representing the major part of energy equivalents.*

Natural gas liquids (NGLs): those hydrocarbons which can be extracted in liquid form from natural gas. Invariably the term NGL embraces propane and all heavier hydrocarbon fractions, ie butane, pentane, etc. In some instances it may be taken to include ethane as well*.

Gas condensate: the heavier NGLs - pentanes (C_5H_{12}) and hexane (C_6H_{14}), which would be liquid at normal temperature and atmospheric pressure, but have low boiling points which cause them to become vaporised and mix with other gases.**

***Proven reserves: those reserves which on the available evidence are virtually certain to be technically and economically producible.

***Probable reserves: those reserves which are estimated to have a better than 50 per cent chance of being technically and economically producible.

***Possible reserves: those reserves which at present are estimated to have a significant but less than 50 per cent chance of being technically and economically producible.

Load curve: A graph in which the send-out of a gas system is plotted against intervals of time.*

Base load supplies: supplies which are required throughout the entire operating period of a gas system.

Peak load supplies: supplies which are required only when demand reaches a daily or seasonal peak.

UKCS: United Kingdom Continental Shelf.

BGC: British Gas Corporation.

* These definitions are taken from Malcolm W. H. Peebles, Evolution of the Gas Industry. London: Macmillan, 1980, pp.211-15.

** Frank Frazer, Gas Prospects in Western Europe. London: Financial Times, 1981, p.9.

*** Department of Energy, Development of the Oil and Gas Resources of the United Kingdom (Brown Book), 1983.

CONVERSION FACTORS

One billion (thousand million) cubic metres (BCM) of natural gas per year is approximately equivalent to:

    0.04 trillion cubic feet per year
    100 million cubic feet per day
    375 million therms per year
    890,000 tons of oil per year
    17,800 barrels of oil per day

Where price conversions have been made between pence per therm and US $ per million British thermal units ($ per mmbtu): 10p per therm is approximately equivalent to $1.60 per mmbtu (where £1 = $1.60).

SUMMARY

The UK currently produces around 75 per cent of its
gas supplies and it is very likely that this level
of self-sufficiency will remain roughly constant
through the early 1990s. After 1995, and
particularly after 2000, the picture inevitably
becomes less clear. Assuming reserves lie at the
higher range of current estimates, it would be
theoretically possible to achieve total self-
sufficiency in the period 1990-2010 and extend this
to 2020 with the addition of four SNG plants using
coal as feedstock. In practice, this would be both
logistically difficult and a most expensive option,
involving a three-fold real cost increase for
conventional gas supplies and a six-fold real
increase for SNG supplies. In the high self-
sufficiency scenario, conventional production falls
to zero after 2020 and the industry is forced to
switch to total reliance on SNG and/or imports.

An alternative scenario would see natural gas
self-sufficiency falling to 60 per cent by 2000,
50 per cent by 2010 and less than 40 per cent by
2020, a level which could be maintained up to the
middle of the century. The gap would be filled by
imports in the short to medium term; SNG would not
be considered an option until the second decade of
the next century, because imported supplies from
most sources are likely to be less expensive. The
low self-sufficiency case would require careful elaboration of a natural gas policy which balances
domestic supplies against imports and weighs the
various sources of imports against each other.
Pipeline sources which are currently attractive are
Norway, Netherlands and the USSR. Large-scale
imports from any of these sources should be combined
with a link to the Continental gas grid which would

improve security of supply for all parties. However, if pipeline supplies from Norway (principally from the Troll field) should prove to be too expensive, and/or supplies from the Soviet Union should be politically unacceptable, the prospects for LNG supplies from Algeria, Nigeria, Canada, Cameroon and Trinidad, could be considered. No LNG supply is likely to be competitive with pipeline imports, except for the highest cost Norwegian gas and the eventual prospect of pipeline supplies from the Middle East. The author's preference is for a gradual slide into greater dependence upon imported pipeline sources of gas. If sufficient reserves are established on the UKCS to maintain higher levels of production than currently expected after 2000, then a higher level of self-sufficiency can be maintained for a longer period of time. Every effort must be made to avoid a sharp discontinuity in sources of supply which could jeopardise the future of the fuel within the energy balance. While greater dependence implies increased vulnerability, there are a large number of supply options from which choices can be made, contingent upon a mixture of security of supply and price considerations. As long as the length of lead times for individual options is appreciated, there is no reason to believe that gas will become a serious area of vulnerability in the nation's energy balance up to 2020.

Introduction

This paper sets out the options for UK natural gas
supply in the period up to the year 2020, with a
view to the degree of self-sufficiency which may be
possible in terms of resources and appropriate in
terms of costs. It reviews the various supply
options which are available, suggesting strategies
for indigenous and imported supply, the decisions
which will need to be made on each of the supply
options, and the timing of such decisions. The
interrelationships between different gas supply
strategies and between gas and other fuels, espec-
ially coal, are considered.

The definitive history of the British gas
industry concludes with the following statement:(1)

> From the outset, before one therm of North Sea
> gas was burned in Britain it was certain that
> the quantity available was limited. How long it
> would last depended on two main factors: firstly
> the total quantity ultimately available and
> secondly the rate at which it was extracted.
> Neither factor was then known with any precis-
> ion, nor is it now: the only certainty was that
> there must be a time limit of some sort. Before
> that limit is reached - or rather, when its
> timing becomes apparent - some new strategy must
> have been developed to ensure continuity of
> supply.

The argument of this paper is that the time limit is
becoming apparent and that it is important to
consider what the elements of that new strategy
might be.

Natural Gas From the UK Continental Shelf (UKCS)

Current UK natural gas production is concentrated
offshore in the North Sea. While other offshore
areas suggest considerable promise for the future,
notably the Irish Sea (where the Morecambe field is
under development), and small finds have been made
onshore, this paper will deal almost exclusively
with the resources of the UK North Sea. Inevitably
there are differences in estimates of gas reserves
on the UKCS and some of these are shown in Table 1.
In each case, the description of the figures is
important to identify, since different organisations
use different classifications; even the UK
Department of Energy 'Brown Book' classification
changes slightly from year to year.(2) Given the
uncertainties, the divergences in Table 1 are
perhaps not as great as might be expected. In the
remaining proven and probable categories, there is
good agreement around the Brown Book estimate of
around 1,000 BCM. The only dissenter appears to be
Phillips Petroleum which sees more than half as much
gas again in these categories. In the more
speculative categories of reserves there is much
wider divergence and a suspicion that the sources
may not be describing the same phenomena. The
Department of Energy figures are naturally the most
cautious, but the British Gas Corporation (BGC), BP
and Shell appear to be very much in the same area;
all of these organisations see a remaining
recoverable reserve figure of 1,200-2,400 BCM.
Phillips Petroleum and Peter Odell both suggest
significantly higher figures of ultimate recover
ability at 3,000-3,500 BCM.

Looking at these round numbers in the context of
the time frame of this paper, with the present level

2

of production of some 38 BCM annually, proven reserves would be sufficient to last until the year 2000 and, assuming ultimately recoverable reserves of 1,400 BCM, the same level of production could be maintained for another 20 years. Of course, these rather simple arithmetical calculations provide only the most rudimentary 'rule of thumb'. It is necessary to look behind the aggregate figures in order to identify the particular fields from which the gas might come. Tables 2 and 3 provide the Department of Energy breakdown of natural gas resources and emphasise the importance of distinguishing between dry gas, associated gas and condensate fields.

Dry gas

Looking at gas reserves from which the vast majority of current UK gas production is drawn, the Southern Basin of the North Sea contains remaining proven reserves of 299 BCM, while the Frigg and Morecambe fields have proven remaining reserves of 169 BCM. Of these 468 BCM, 420 BCM are already 'in production or under development'. It appears that at the current rate of production (see Table 4) the proven reserves of the Southern Basin fields are likely to be exhausted within eleven years. Even this approximation is misleading, however, since some fields such as Rough will be used for peak supply, thus curtailing further the period of uninterrupted base load supply from the fields. Probable and possible reserves from the Southern Basin would extend production of the fields by an additional five years at current levels.

The reason for concentrating on the Southern Basin in some detail is that these are the fields which carried the UK into the natural gas era. Of the other gas fields which have been located thus far on the UKCS, only Frigg is comparable in size with the Southern Basin fields, although the field is far less conveniently located (expecially since 60 per cent of the gas is under Norwegian jurisdiction). Other fields which may be discovered in the future are likely not only to be of smaller size (and probably more complex geologically), but also geographically situated further from the mainland, requiring long undersea pipelines to shore. The point, therefore, given the resource base thus far identified, is that, as the Southern

3

Basin becomes exhausted over the next 10-15 years, replacement reserves will be more difficult and expensive to develop.

Of the other dry gas fields, the UK section of Frigg field, with some 60 BCM of remaining recoverable reserves, will be nearly exhausted by the early 1990s. The Morecambe field in the Irish Sea, containing some 140 BCM of proven reserves, will have the capacity from the mid-1980s to deliver a minimum of 600 million cubic feet per day (mcf/d) - equivalent to a yearly delivery of 6 BCM - and a maximum of 1,200 mcf/d by the end of the decade.(3) It is planned, however, to release these volumes only at the time of maximum demand during the winter months. Thus it is not appropriate to put an estimate on the life of Morecambe reserves, or to use these reserves when calculating the base load supply availability, although it would be possible to use Morecambe gas in this role in the future, if assumptions were to change on the most profitable mode of utilisation. Aside from probable and possible reserves in the Frigg and Morecambe fields, there is only another 42 BCM of possible reserves of dry gas on the UKCS. The remainder of the reserve base is to be found either in condensate fields or in fields where the gas is associated with oil.

Associated gas and gas condensate.

Remaining proven reserves of associated gas amount to 125 BCM, with 96 BCM of probable and possible reserves in equal quantities. Natural gas reserves in gas condensate fields amount to nearly 200 BCM proven and probable (overwhelmingly concentrated in the latter category) reserves and 337 BCM possible reserves. On these calculations, one quarter to one third of the remaining recoverable UK gas reserve is in oil and/or condensate fields. The majority (and given the inability to forecast future condensate discoveries, perhaps the vast majority) is in condensate fields and a very large part of the reserve is in the probable and possible categories.

Natural gas found in association with oil and natural gas from condensate fields gives rise to a number of technical and organisational problems, related to treatment and transportation, particularly when deposits are situated offshore in geographically dispersed locations. What is required is a

4

number of pipelines connecting oil fields and condensate deposits, and feeding into a main gas trunkline to shore. This is generally referred to as a 'gas gathering pipeline' system.

Where gas produced in association with oil cannot be reinjected into the deposit, or collected and piped to shore, it must be flared if oil production is to continue. Flaring of associated gas is a problem for oil producers worldwide; up to 1980, OPEC countries had been flaring more gas than they utilised.(4) The volumes of flared gas have been substantially reduced by the construction of huge gas gathering networks requiring massive investments; the Saudi Master Gas System, which can cope with the gas associated with a production of some 7 million barrels of oil per day, was completed in 1982 at a cost of $14 billion.(5) It is probably fair to say that, only following the 1973 world oil price rise and the consequent rise in natural gas prices, has much of the associated gas worldwide become economic to gather. Even now, many would argue that the gas marketed by the Saudi system will not realise an acceptable commercial rate of return on such a large investment.

While it might be possible for countries with small populations and low (or zero) domestic demand for gas to contemplate substantial flaring, for the UK, with a very substantial domestic demand (partly sustained by large-scale imports), the flaring of more than marginal quantities of fuel must be a matter of serious concern. During the period of large-scale oil production (1977-82) on the UKCS, some 28 BCM of gas was flared, equivalent to about 12 per cent of production over the period. Volumes of gas flared rose to a peak of 6.7 BCM in 1979 (Table 5), of which one half came from the Brent field where, as a result of technical problems and newly imposed government flaring restrictions, the operator was forced to make a temporary cut in the rate of oil production.(6) The likelihood of increased flaring, combined with the advent of sharply increased oil prices (making economic a number of fields hitherto considered marginal) and an anticipated shortage of energy and natural gas, gave rise to an appraisal by the British Government of the different gas gathering options in the UK North Sea.(7) The option which received most serious consideration was the joint BGC/Mobil ' Northern

North Sea Gas Gathering System', designed to transport natural gas and natural gas liquids (NGLs) from both the UK and Norwegian sectors to the UK mainland. The BGC/Mobil study concluded that their system would yield approximately 11 BCM of UK gas per year in the late 1980s and early 1990s, augmented by roughly the same quantity of Norwegian gas. The total quantity of UK gas available for collection was estimated to be some 150 BCM, in addition to which some 42 million tonnes of NGLs would be made available (3.4 mt in a peak year), which might be augmented in the 1990s to a total recoverable reserve in excess of 93 mt.(8)

The reasons why the system was eventually abandoned are complex, and different actors will lay the stress and/or the blame on a whole range of factors and third parties. Certainly a considerable setback was the loss of Norwegian Statfjord gas to the Continental countries, and there are many who believe that in the absence of Norwegian gas the system was economically unviable. However, even without this problem, a vicious circle of attitudes had developed. The financial community would not lend funds - the estimate of which had risen from £1.1 billion to £2.7 billion - without a throughput contract in existence between the oil companies and BGC, or a guarantee of repayment from the oil companies. In turn, the companies would not commit funds without a commitment from BGC on the price of the gas. BGC would not give a commitment on price without a volume guarantee from the companies. The logjam created by these attitudes threw the burden of financing back on the government.

The Conservative Government had intended the gas gathering systems to be operated as a privately-owned public utility, with the oil companies as the major shareholders along with BGC. A significant practical difficulty with this concept was the fact that such an entity would have had no immediate profits against which to offset its costs and indeed would not have been able to see any such profits for some years to come. More substantial State financing, which had become the only possible option for keeping the project alive, was unacceptable to the Government which took the view that, since the private sector had refused to finance the project, the companies should 'make their own arrangements for bringing the gas ashore. It is confident that

... the producer companies will ensure that Britain's North Sea gas reserves are brought ashore efficiently in accordance with the nation's needs'.(9)

In effect, this left the Shell/Esso Far North Liquids and Associated Gas Systems (FLAGS) as the sole gathering facility for UKCS gas. Originally designed to transport the associated gas from the Brent field to St Fergus, a 'western leg' including gas from the North and South Cormorant, Ninian and N.W. Hutton fields was added to the plans in 1978. A 'northern leg' was acquired in April 1981 when the gas from the Magnus, Murchison and Thistle oil fields was switched from the proposed gas gathering system (even before that scheme was abandoned). (10) At the beginning of 1983, there was speculation that the gas from the UK portion of the Statfjord field would be delivered via the FLAGS line, the UK Government and BGC having refused a request from Statoil (the Norwegian State oil company) to allow the UK gas to be delivered to the Continent via the Norwegian Statpipe system.(11) There are, however, indications that UK Statfjord gas was used as a bargaining counter (albeit a rather small one) in the negotiations to bring Norwegian Sleipner gas to the UK (see below). The FLAGS line was commissioned in 1982 and is being raised to full capacity by the end of 1985. This should considerably reduce the quantity of gas being flared from oil fields currently in production. Other fields from which associated gas is being currently collected are Piper and Tartan from which gas is fed into the Frigg line; there are plans to pipe gas from the North Alwyn field to join up with the Frigg line. In addition, a line is being built from the Fulmar field to St Fergus, which should carry gas and NGLs from Fulmar, Clyde and probably a number of other small condensate fields, starting in 1986. Initial volumes of gas will be small, probably less than 1 BCM per year.(12) Thus the main issue remaining from the large-scale gas gathering system is the means of transportation for UK Statfjord gas, which probably cannot be transported via FLAGS before the early 1990s without 'backing out' some of the Brent gas.(13)

The collapse of the BGC/Mobil system raises a number of interesting questions about the organisation and financing of future gas gathering efforts.

7

There is a strong argument that a gathering system needs a 'lead field' - a large gas source to provide the backbone of the line and the focus for others to be drawn in: Brent in the case of FLAGS and (Norwegian) Statfjord in the case of BGC/Mobil. Whether or not this is so, it seems clear that in the absence of a buyer who is prepared to sign up a number of sources in advance, with some assurance on prices, private funding will be impossible to arrange. This leaves the option of direct State financing, which seems unlikely to appeal to a government which has previously rejected such an arrangement.

One part of the current emphasis on the organisation of gas gathering is therefore on private rather than State-owned schemes; the other part is on small-scale rather than large-scale systems. In practice, this is likely to mean that one operator has to take the initiative to build a line from a particular field to shore with some built-in spare capacity to allow for others to feed in at a later stage, rather than a large-scale private or State-owned system. Both alternatives have drawbacks: the small-scale system will only encompass those fields which look certain to yield an acceptable commercial return on the investment within a short period of time. Operators are more likely to want to put gas from their own fields through their own pipeline, rather than through that of a competitor. Spare capacity may be too expensive to build into a pipeline unless a future partner is willing to put up some investment ahead of time, or the fiscal regime makes some allowances for this.

On present indications, therefore, a great deal less gas will be gathered in the short term by small schemes than would have been the case under the much larger BGC/Mobil system.(14) This may not matter if the gas is gathered at a later stage, but there is the danger that, under pressure to maintain and/or increase oil production, a proportion of associated gas will be flared which might otherwise have been collected and used. The situation will depend on the strength of the government commitment to a tight gas flaring policy.(15) In turn, this commitment will be affected by the degreee of government reliance on a high level of crude oil production and exports for the raising of revenue. The greater the dependence on high oil revenues, the greater the temptation to permit a high level of gas flaring.

In drawing up a large system, there is the problem of ensuring that the individual field operators can co-ordinate development in order to dovetail with throughput availability in the system. There must be a question as to whether systems which involve a large number of operators are workable, even with maximum co-operation and goodwill on the part of those involved. The sheer complexity of negotiations between a large number of partners means that the lead times for reaching agreement to construct the system are likely to be extremely lengthy - probably longer than the actual physical construction of the system itself. In addition, a system which is designed to include a large number of fields probably has to take an optimistic view of (oil and) gas prices and hence the economic viability of fields which appear marginal at the planning stage. If the assumptions have not been borne out by the time of construction, the choice may be to produce gas at an uneconomic cost or to run the system well below capacity. There is also the problem of ensuring that the operator of each field receives the correct financial return on the gas which is contributed to the total system. This problem is common to all gas gathering systems, but is compounded when the number of fields from the which gas and NGLs are being collected is very large.

Condensate fields raise other complex problems, in addition to the gas gathering and transportation issues. In many instances there will be a need to recycle the natural gas if the NGLs are to be produced to the fullest extent, and recycling may continue for a period of several years before the natural gas can be produced. The alternative option is to produce the natural gas at an early stage and forego the NGLs. A decision of this kind would probably require government permission.

There are thus no simple answers to the problems of gas gathering, and it is clear that the UK is some considerable way from being able to create access to the 370 BCM of remaining proven and probable associated gas and condensate reserves, let alone the same quantity of possible reserves which are thought to be available. When the FLAGS line is

fully commissioned the average annual capacity of
associated natural gas pipelines to the UK mainland
will be around 7 BCM, (6 BCM via FLAGS and up to an
additional 1 BCM through the Frigg line).(16)   In
the early 1990s spare capacity will open up in the
Frigg line, which may solve some problems for fields
in convenient geographical locations.   Other
possibilities would be opened up by additional
pipelines bringing in imported gas. In the event of
the UK contracting to import Norwegian Sleipner gas
through a dedicated pipeline (rather than through
the existing Frigg line), there would be an
opportunity to design the route of the line to pick
up gas from a number of oil and condensate fields.

The reason for discussing associated gas and
condensate prospects in some detail here is that
these are critical and uncertain elements in total
gas production, the future resolution of which will
have important repercussions on gas selfsufficiency.
Gas which is flared is lost permanently from
reserves, while gas which has been located in a
deposit far from existing pipelines may be many
years away from being available for consumption.
Indeed, the UK may have to face the possibility that
the development of gas in some fields will not be
considered economic for a considerable period of
time, perhaps never.

To sum up, the lack of a cost curve for UK
natural gas and condensate fields is an enormous
handicap in assessing potential production rates.
Another major difficulty lies in the assessment of
lead times necessary for production, particularly of
associated gas and condensate fields. When 'new'
(higher) gas reserve estimates for the UKCS are
announced, it is always worth asking what has
changed in order to give rise to these higher
figures:
  a) new discoveries which have led to an
      increase in producible reserves in place;
  b) a more favourable tax regime which
      encourages companies to develop reserves
      identified some time previously. This
      should be distinguished from a more
      favourable tax regime, and/or a promise of
      higher gas prices, which may encourage
      greater exploration for reserves which <u>are
      believed to exist</u>, but about which little is
      known.

10

Once it has been established that new reserves -as opposed to the promise of new reserves - have been discovered or created by an improved fiscal climate, the question is how long it will take to produce them. For example, there are those who think that some recent, small finds in the Southern Basin could make a significant impact within a ten-year time frame; for new finds in the Central Basin, the time frame would tend to be more than ten years.

The future economic viability of dry and associated gas resources depends on the cost of production, the price offered for the resource, and the government tax policy. The quantity of gas available at any given date will be a function of how operators see the levels and interplay of those three factors currently and also five to ten years hence. The long lead times are worth stressing; the first surveys were made for the FLAGS line in 1975 and it will be fully operational only in 1985.(17) Such factors must constantly be kept in mind when discussing reserve levels and plans for future production being made by the private sector. The decisions of a State-owned utility may be somewhat different in respect of costs of production and rate of depletion of the resource. Uncertainty sur-rounding the size and cost of future domestic gas supplies may therefore, paradoxically, be greater than for certain external sources of supply.

However, from a purely resource perspective there appears to be no problem of maintaining prod-uction through the mid to late 1980s at current levels. In the early 1990s, both the Southern Basin fields currently in production and the Frigg supply (both UK and Norwegian) will start to decline, and a gap will begin to open up. There is support for the thesis that, on purely reserve considerations, there is no reason why this gap should not be filled entirely from UKCS resources; but there must be considerable doubts that this will actually occur. The reason for these doubts relates to the lack of discoveries of large accumulations to replace both the Southern Basin fields and Frigg, combined with the lack of a large-scale gas gathering system of the type contemplated under the original BGC/Mobil system. In order to offset these trends (and assuming the present level of reserves), smaller accumulations of dry gas, associated gas and

11

condensate fields will need to be developed, which will require long lead times and a complex network of gas gathering lines. This undoubtedly suggests a much higher cost of production and transportation for gas from future deposits; just how much higher may be the critical question for future production levels from the UKCS.

The UKCS is not the only option for meeting future demand, however. Imports of natural gas from a number of sources could be expanded and it is to the various sources of availability outside the UK that we now turn.

External Sources of Natural Gas

Pipeline gas from Norway

Gas imports from Norway began in 1978 through a
pipeline from the Norwegian sector of the Frigg
field to St Fergus. The UK had earlier bid for the
gas from the Norwegian Ekofisk field, but had lost
out to the Continental European consortium of
utilities from West Germany, France, Belgium and the
Netherlands. The disappointment of losing Ekofisk
gas was compensated by the contract for more than
149 BCM of Frigg gas. With some 43 BCM already
delivered in the 1978-82 period, the contract allows
for around 10-11 BCM of gas to be delivered annually
up to 1990, with volumes running down to zero by
1993(18).

At the end of February 1984, BGC had completed a
contract with Statoil for gas from the Sleipner
field.* Sleipner is estimated to contain up to 200
BCM of recoverable reserves in four blocks.(19) Most
projections of field developments suggest a possible
start-up date in the early 1990s, with a plateau
production in the mid-1990s of 11 BCM per year;
optimistic forecasts speak of starting in 1989 with
a plateau production of 13 BCM per year in the early
1990s. In any event, the timing and volumes would
mesh well with the expiry of the Frigg contract;
Sleipner would provide an almost direct replacement
for (at least the Norwegian portion of) Frigg gas.

Later in the 1990s, the UK is likely to be
seeking additional supplies from Norway, whether or
not Sleipner is secured. At the present time, these
supplies appear likely to come from the Troll field.

* This contract had not yet been approved by the UK
Government.

Delineation of the field is not yet complete, but the indications are of at least 1,600 BCM of gas (and a considerable quantity of oil) in a very difficult geological formation at a water depth of 350 meters.(20)   Reserves in the Troll field are thought to be so large that, with an unconstrained level of production, competition with Continental importers should not be a problem.   This would eliminate a major obstacle encountered thus far in bidding for Norwegian gas supplies.   The problem in the case of Troll is likely to be the cost of production and hence the price that Norway will demand in order to develop the field.   On present plans, development is to be in two phases: a first phase, which seems likely to be a high oil/low gas production development, will start in the early to mid-1990s; the timing of the second phase – which would produce appreciable volumes of gas – is uncertain and may depend on the success of the initial development.(21)   However, there seems no reason why Norway could not be supplying the UK with at least 20 BCM of Troll gas by the middle of the first decade of the next century.

Supplies from Norway will arrive by pipeline, although the route of such lines is uncertain and, as indicated above, may be determined by the development of gas fields on the UKCS.   It may be possible (and certainly less costly) to use existing lines, ie Frigg and Ekofisk, as those facilities become idle with the exhaustion of resources. Another interesting prospect is the creation of a new pipeline system through the UK to be used as a transit route for Norwegian gas to Continental Europe; this will be discussed below.

Pipeline gas supplies via Continental Europe

The prospect of pipeline gas from the Netherlands was raised in the early 1960s, after the discovery of the Groningen field and prior to the finds on the UKCS.(22)   With the discovery of the Southern Basin fields, these ideas were abandoned and exports of Dutch gas were concentrated in the Continental market.   With Dutch gas largely committed to export markets and the country already importing gas from Norway, the prospect of exports to the UK would now seem unlikely, were it not for the apparent change in Dutch export policy in the 1980s, consequent upon a reappraisal of reserves.   The Netherlands has

14

discovered new reserves and uprated existing fields, raising the prospect of prolonging large-scale export contracts beyond the end of the century when they are due to expire.(23) Alternatively, the Netherlands has been suggested as the primary 'surge supplier' or supplier of last resort for Continental countries which may become exposed to supply interruptions during the 1990s and beyond, as a greater proportion of their gas comes from non-European countries, namely the USSR, Algeria and elsewhere.

The UK might welcome the prospect of Dutch supplies, either on a large scale or as a peak load facility, particularly as the Southern Basin fields become depleted. The logistical link between the Dutch and UK North Sea sectors would be short and might not require a great deal of additional pipeline capacity. Although supplies from the Netherlands would probably yield only around 5 BCM per year, there were signs, at the end of February 1984, that this was being seriously considered by the UK Government as an alternative to large scale Norwegian Sleipner gas supplies. Significant peak load capacity could be built into a line from the Netherlands.

Possible complicating factors to imports of Dutch gas would be the fact that the cost of the link to the Southern Basin would undoubtedly have to be borne by the UK and might not be considered worthwhile if the quantities were to be small. In addition, there would be implacable opposition from the Continental importers to the idea that Dutch gas should be exported anywhere other than to their market. It is likely that at least some export contracts to Italy and France will not be renewed when they expire. This loss will be keenly felt by importers who will not look kindly upon new supplies to the UK starting, seemingly at their expense, when the UK resource base gives it such a relatively comfortable position compared to its Continental neighbours.

The USSR already supplies five West European countries with natural gas: West Germany, Italy, Austria, Finland and France (in addition to six East European countries and Yugoslavia) through a grid which stretches as far west as Paris. A link to the UK would require an extension of the pipeline network westward and a cross-Channel pipeline to the

UK mainland, or conceivably a displacement agreement with the Netherlands (this will be discussed below). It is clear that the USSR, with 40 per cent of the world's proven gas reserves, has a great deal of natural gas to sell to Western Europe and is extremely keen to do so in order to earn hard currency.(24)   Unlike all other external sources of supply, the timing of Soviet deliveries would be affected only by problems of logistics.   The USSR would be in a position to supply very large quantities of gas to the UK within a comparatively short time.   With the building of the new Urengoy pipeline it would be possible for some 10 BCM of Soviet gas per year to be flowing to the UK by the early 1990s.(25)   The only physical (as opposed to political) constraint is the speed at which the cross-Channel link could be built.   In theory Soviet gas could have been in competition with Sleipner gas for the UK market in the early 1990s.   In any event it would be a strong competitor in the UK choices for gas supplies in the mid to late 1990s and the early part of the next century.

The UK market has a considerable attraction for the USSR, since it is the only major gas market in Western Europe (with the exception of the Netherlands) which it has not yet penetrated.   For the UK, Soviet gas may be an attractive commercial proposition.   It is obvious that the Soviet authorities wish to sell large quantities of gas in Western Europe and are being prevented from boosting their supplies to existing customers by concerns about security of supply, which were voiced in the trans-Atlantic debate over the new Siberian pipeline.   The UK market is sufficiently large for security concerns to be less acute; even if the USSR were to supply 10 BCM per year to the UK by the end of the century, this would amount to, at most, 25 per cent of total supply.   Beyond the end of the century there would be scope for increasing imports from the USSR since there is every likelihood that sufficient export capacity will be made available in the longer term.   The UK could increase its imports of Soviet gas to around 15-20 BCM before the informal agreement between NATO countries, which would prevent gas supplies from the USSR exceeding 30 per cent of total consumption, would come into question.(26)   The actual figure would depend on the capacity of the cross-Channel line; the decision on the dimensions of the facility would therefore be an important consideration.

Other suppliers of pipeline gas to the UK in the future might be North and West African countries and, eventually, the Middle East and Gulf countries. A potential contributor to a cross-Channel pipeline might be Algeria, which has recently begun to pipe natural gas across the Mediterranean to Italy and where feasibility studies have confirmed the possibility of laying a pipeline across the Straits of Gibraltar to Spain. Given the possibilities of pipeline gas from Siberia to the UK, there is no logistical reason why Algerian gas should not travel the considerably shorter distance to the UK via Italy or Spain. The major obstacle at present is Algerian pricing policy which has complicated the trade with Italy and the LNG trade with France. Nevertheless, it would be wrong to rule out the long-term prospects for Algerian gas exports to the UK. The country's resource base and current export commitments would seem to indicate that a long-term export commitment of 5-10 BCM per year could be allotted to the UK. The lower end of the range would be more realistic, however, and it would therefore be reasonable to conclude that a cross-Channel pipeline would not be worthwhile to construct solely for Algerian gas imports.

Similarly it is reasonable to consider imports of pipeline gas from as far away as the Middle East and West Africa. These sources are usually considered only in connection with trade in liquefied natural gas, but if a cross-Channel pipeline were to be in existence, there would be an attraction in capitalising on as many sources of gas as possible. Early in the next century, Middle East gas may be considered more carefully by Continental European countries, given the vast volumes of natural gas in the region which appear to have no alternative large-scale domestic or export market. Pipelines from the Middle East to Europe (via a number of possible routes) or from West Africa (mainly from Nigeria but also possibly Cameroon) via a Trans-Saharan pipeline from Nigeria to Algeria and thence to Europe are technically feasible, though the cost of transportation would be an important element affecting their economic viability. However, the main obstacle would be the number of countries the lines would have to cross, which would require detailed security analysis. The Continental countries are likely to lead in this endeavour, but the UK should certainly take an interest in receiving a share of the very large volumes which would need to be involved in order to make the projects economically viable (particularly if British companies were to be leading actors in proposing, designing and constructing such pipelines).

A cross-Channel pipeline: exports, imports, displacement agreements and security

There has been very little public discussion of the possibility of a cross-Channel gas pipeline. In the 1978 Green Paper it was noted:(27)

> If the UK and Norway decided to collaborate in the development of gas resources in the Northern Basin of the North Sea, it might then be feasible, if the Norwegians so wished, to re-export their portion of the gas across the Channel to the Continent. In the long term, when our supplies are in decline, a cross-Channel link would open up the possibility of the UK importing gas by pipeline from Asia or the Middle East.

There was reiteration of this proposal to the Norwegians at the time of the Strafjord negotiations. However, there are a number of different alternatives raised by the prospect of a cross-Channel pipeline. In addition, there is the prospect that the UK might play a significant role in an interconnected West European gas grid which might enhance security arrangements for the entire region.

During the 1970s, there was opposition from both the government and BGC to the idea of a cross-Channel pipeline or any other pipeline which might possibly allow UKCS gas to be landed anywhere other than the UK mainland. Given the statutory responsibilities of BGC to the UK consumer, this is entirely understandable. Internationally traded gas is sold on long-term contracts and any trade which would warrant the building of a pipeline would involve rather large volumes. From the reserve position detailed above, it is obvious that a contract which involved even 5 BCM per year (itself

18

hardly enough to warrant construction of an undersea pipeline) for twenty years, thereby subtracting 100 BCM from the proven reserve - nearly 16 per cent of the present total - could hardly be considered in the national interest. It is equally understandable that oil companies, which consider that they have always been paid too low a price for their gas by BGC, wish to gain access to a market which has a tradition of paying higher gas prices, particularly in cases where the pipeline link to the Continent would be easier than to the UK.

A BGC view expressed to a House of Lords committee in mid-1982 was that:(28)

> We believe at the moment that the establishment of a pipeline link very clearly carries with it the implication of the export of indigenous gas to Europe and we express concern about that suggestion.

Such fears were heightened earlier in the year when the Secretary of State for Energy raised the prospect of exporting gas from the UKCS. At the same time as extolling the virtues of the Oil and Gas Enterprise Bill (1982), he added:(29)

> all the gas in our offshore fields currently in production is contracted to be sold in Britain, so the question (of exports) arises only for future fields ... If, however, the fresh impetus which our policies will undoubtedly give to exploration, results in large volumes of new gas being discovered, the question of exports can and will be reconsidered then.

He did not make it clear how much gas would need to be discovered in order to raise the possibility of exports and whether he was thinking of large-scale quantities or marginal volumes. It would be difficult to predict the quantity of gas which would need to be discovered to allow exports of more than marginal volumes, while not endangering long-run security of supply. An example of the pitfalls which lie ahead in any such exercise can be seen in the case of the Netherlands, where large-scale exports were contracted in the 1960s, only for the jolt of the 1973/4 oil crisis to bring the country to contract for imports of Norwegian gas and then subsequently in the 1980s to reverse its policy again by raising the prospect of increased exports. This provides an example of how a country, with a smaller domestic consumption than the UK and more than twice the proven reserve base, can be panicked into taking illogical action when its long-term security of supply appears to be threatened. For the UK to contemplate large-scale exports when already importing 20-25 per cent of its gas supplies would be an evident absurdity.

Although large-scale exports of UK gas may be out of the question, there are two areas which warrant rather more careful scrutiny. The first concerns co-operation with Norway in the exploitation of the associated gas in fields which are close to, or straddling, the sector line. For example, it is evident that UK Statfjord gas could be more easily transported through the Statpipe system to the Continent (once the decision had been taken to send the gas from the Norwegian sector to that destination) than to the UK mainland. It would be possible, as regards UK Statfjord and other fields close to the sector line, to be flexible about the destination of the gas, provided that a quid pro quo could be worked out with base load Norwegian supplies. Thus the idea of using small quantities of UK gas as a bargaining counter in negotiations with Norway should be acceptable.

The second area where exports of UK gas should be seriously considered concerns the concept of West European gas security and the future involvement of the UK in the Continental gas grid. There is no counterpart in natural gas trade to UK participation in the IEA emergency oil sharing mechanism, or the cross-Channel electricity link (with France) which will commence in 1985.

It would be of considerable benefit to Continental countries to have the assurance of UK supplies for emergency contingencies. The French company Elf Aquitaine expressed it thus:(30)

The general interconnection of gas pipelines in Europe would allow full benefit to be gained from the application of measures designed to reinforce security of supply. With regard to a connection between the UK and the Continent, a pipeline the main purpose of which would be solely designed to equalise occasional variations in supply would, according to the (European) Commission, 'be difficult to justify on the basis of the analysis of cost and advantages', but if there were gas to be transported, this would certainly be of interest to the Continent. ... A scheme of this nature would normally receive assistance from Community organisations, particularly in the form of special financial arrangements.

However, it is difficult to imagine that a pipeline to the Continent, the sole purpose of which was to permit occasional gas transfers for security reasons, would be possible to finance commercially or of any concrete benefit to the UK. To the sceptical, this would still seem a thinly disguised excuse for exports of UK gas on a more than marginal scale.

By contrast, it is perfectly possible to envisage a pipeline which primarily involves imports of gas, but provides for exports in emergency situations. The most obvious example would be large-scale imports of Soviet and/or Algerian gas through a cross-Channel pipeline from either France or Belgium. Another would be the re-export of Norwegian gas transitting the UK, which might involve an export of UK gas from the south of England as a quid pro quo for additional imports of Norwegian gas rather than exports of UK supplies. Yet another example would involve the UK receiving Soviet gas in a displacement agreement with the Netherlands. In this scenario, the Netherlands would redirect contracted supplies away from Continental countries which would receive additional supplies of Soviet gas. With the transportation capacity remaining intact, the Netherlands would retain the ability to switch supplies back to Continental countries if an emergency were to arise; additional UK gas might then be required to reinforce Dutch supplies. In early 1984, a load balancing arrangement was under discussion between the Netherlands and the UK.

In each case the only additional requirement would be the facility to reverse the flow in the line when security dictated. Alternatively, in the case of the link to France or Belgium, it would be possible to lay two pipelines side by side enabling two-way flows of gas to take place. This latter

21

option would be more expensive but might be easier from a contractual standpoint, since the normal method of fixing a minimum yearly delivery volume of gas through a pipeline might not be susceptible to sudden changes in the direction of flow.

Quite aside from any commitment to a 'European energy (or gas) policy' which may be implied by membership of the EEC, the UK will certainly need the goodwill of Continental countries if gas destined for the UK is passing over their territories or through their facilities. The adversarial relationship between BGC and the Continental utilities in bidding against each other for the same sources of supply may fade in the future. It could be advantageous for all parties if this were to be replaced by joint co-operation in expanding the supply flexibility. A cross-Channel pipeline would be a considerable asset in this regard.

LNG supplies

The first commercial liquefied natural gas trade was initiated between the UK and Algeria in 1964. A regasification terminal was constructed at Canvey Island and this handled about 1 BCM of LNG per year from the Arzew 1 liquefaction plant until the contract expired in 1980. A supplementary nine-month agreement expired in October 1981. Since that time no LNG has been received at Canvey Island, partly because of an accident which severely damaged the jetty in May 1982.(31) This would in any case have precluded the berthing of LNG tankers, but there has been no suggestion of restarting trade with Algeria in the interim period; the jetty was repaired by the autumn of 1983.

By modern standards, the Canvey Island terminal is extremely small and can only be considered a 'peak load' facility. Quite different facilities would be required if LNG supplies were to take over a 'base load' role in the UK. The most recent LNG terminals built in France and Japan have a capacity of 7-12 BCM per year and the UK would require 2-4 terminals of this size to meet 50 per cent of present demand.(32) Furthermore, the lead times for constructing a modern LNG terminal are very considerable. BGC estimates the construction time for a terminal as:(33)

at least five years, depending on the amount of site preparation necessary. The award date would be preceded by a period of time during which conceptual designs would be prepared, including hazard analysis and environmental impact analysis studies and planning applications made. The minimum time for this activity would be two years. However, this preconstruction period could easily be extended if a public inquiry was to be held and an extended period of public consultation necessary. In this event, the preconstruction period would last 4-6 years. Thus the minimum time needed to bring an LNG terminal to the operational stage from initial nomination of a specific site would be 7 years, but this could be extended to over 10 years.

In preparing this estimate, BGC may have in mind not simply the experience of other countries, but the potential opposition from local residents which, in the case of Canvey Island, has led to three reports on the safety aspects of the facility since 1978.(34) The most recent report (1983) concluded that, after a period of unsatisfactory performance in the late 1970s, the terminal could be considered safe and that capacity could be maintained at current levels. However, BGC gave an undertaking that the capacity of the terminal would not be expanded. Thus new sites would be required in order that large-scale LNG supplies could be imported. Using the time frame outlined above, this would preclude any large-scale import of LNG before the early to mid-1990s, even if a terminal were under consideration by BGC; there is no present indication that this is the case.

As regards sources of supply, Algeria would certainly be the prime source of LNG (presupposing that this was considered preferable to the pipeline alternative outlined earlier). Other possible sources of LNG would include Nigeria, (Arctic) Canada, Trinidad, Cameroon and the Gulf (principally Qatar). On present knowledge, it would be unrealistic to assume that Canada and Trinidad would yield more than marginal volumes of gas in any time frame. For domestic economic and political reasons neither Nigeria nor Qatar is likely to develop a transportation infrastructure which would yield large quantities of gas before 2000, quite apart from the question of UK readiness to accept such supplies.(35)

After 2000 and certainly prior to 2020, it is perfectly feasible that Nigeria, and particularly the Middle East countries, could be sources of considerable volumes of LNG (assuming they had rejected the pipeline alternative). However, there are considerable uncertainties, particularly with regard to the route for the Middle East trade, where it is not certain that large numbers of tankers passing through the Suez Canal would be environmentally acceptable; the route around the Cape would be enormously costly. The economics of Nigerian LNG are also uncertain, primarily because of the large investment required for gas gathering and liquefaction installations in the country. In the future, it is quite likely that countries with large populations may find attractive domestic uses for gas at the expense of exports.

Substitute Natural Gas (SNG)

Aside from natural gas from the UKCS and the various import options, there is the prospect of producing gas from other fuels - mainly coal and oil. This is generally known as SNG or 'substitute' natural gas, and there are a number of different technologies which are in various stages of commerciality. At the present time, there is no large-scale high calorific gasification programme in existence, and many West European countries have recently shelved plans for starting such a programme because of falling oil prices. The only consideration here will be high calorific value gasification, ie gas which is interchangeable with natural gas in the national grid, and specifically, gasification of coal. There are also options to produce lower calorific value gas from coal, which can be used in combined cycle power plants or as synthesis gas. As far as gasification of oil and oil products (heavy oil and naphtha) is concerned, although the UK may run short of natural gas from the continental shelf prior to exhausting its oil reserves, the cost of production and the international price of oil are likely to preclude large-scale conversion from oil to gas. Even so, the two plants at Portsmouth and Plymouth which use naphtha as feedstock are an indication that BGC is interested in this prospect. The primary concern must be to avoid the conversion of any part of the gas grid to lower calorific value gas. Having made the transition to natural gas with a massive conversion effort in the late 1960s and early 1970s, it would be inconvenient (as well as expensive) to reverse that process in the future.

BGC has been conducting a long-term research programme of coal gasification at Westfield in

Scotland. In 1981, the BGC/Lurgi slagging gasifier was reported to have completed 90 days of extended trials during which 1,750 cubic feet (approximately 50 cubic metres) was gasified from 27,000 tons of coal at a rate of 300 tons per day.(36) Commercial SNG plants are likely to produce 2.5 BCM per year, requiring an average of 5 mt of coal per year (this is an approximate figure which ranges from 11,500 tons per day high quality coal to 25,000 tons per day for poor quality lignite). At present, 20-25 plants would be needed to satisfy the whole of UK gas demand, and this would require 100 mt of coal per annum. The approximate capital cost is thought to be around £1 bn per plant and the cost of the gas is estimated to be 'two to three times as expensive as natural gas', or 'upwards of $50 per barrel (1982 dollars)'.(37)

The sites for SNG plants require access to large quantities of coal and water (50,000 cubic metres per day) and disposal facilities for solid waste. A commercial gasifier is estimated to produce 2,000 tons of slag, 230 tons of sulphur and 5-10 tons of sludge on a daily basis.(38) The Commission on Energy and the Environment suggested that there was a need for an intermediate demonstration plant of 0.5 BCM per year capacity, prior to the building of full scale 2.5 BCM per year plants.(39) BGC estimates that:(40)

> from the date of award of the design and construction contract to satisfactory operation, a period of 7 years would elapse. As in the case of any major industrial site ... the preconstruction period could be considerably extended by the requirements of public planning inquiries. The whole process could therefore well extend to 10-14 years. Based on the construction period alone, it is estimated that a period of up to 10 years would be needed to create 3,000 million therms (about 8.3 BCM) of SNG and up to 15 years to create 6,000/9,000 million therms (16.5-25.0 BCM) of SNG capacity, following the successful completion of a proto-type plant. From about 1985, when further development work has been completed, it would be possible to embark upon a construction programme which would give SNG by the mid-1990s if this were essential. This would imply a bypassing or modification of the prototype programme and

could give rise to increased costs of construction and operation.

A report in 1983 that BGC was intending to build 20 coal gasification plants, starting in 1990, would therefore imply an accelerated programme, either speeding up or eliminating the prototype stage.(41) Even starting in 1990, the first plants would probably not be ready much before 2000-2005 and it seems likely that one would be built and tested before a full-scale commitment of funds was made. Alternatively, construction and testing of a prototype plant might push the first commercial station back to the second decade of the next century. In any case it would seem prudent not to rely on large-scale supplies of SNG before this date. The only development which could radically change these judgements would be a technological advance in the United States, leading to rapid commercialisation, which could be imported by this country on a large scale. While this should not be ruled out, the current slump in both interest in, and funding for, synthetic energy technologies in the United States does not give cause for optimism.

Overall, it is clear that the lead times and the technical uncertainties are such, that to produce appreciable quantities of SNG prior to 2000 would require a very large commitment of funds and would also envisage moving ahead without a properly tested technology at the prototype stage. This is important since SNG from coal may provide the only domestically produced gas available after the depletion of the UKCS. SNG is therefore crucial to any aspiration towards natural gas self-sufficiency. It would, of course, be possible to base an SNG programme on imported coal, although it is easy to envisage tremendous resistance from the British coal industry to such action. SNG production from imported coal would not strictly represent self-sufficiency in fuels, however, and it would be unlikely that this course would be adopted unless there were a substantial price advantage.

Costs and Prices of Domestic and Imported Natural Gas

Perhaps more than any other issue pertaining to
natural gas supplies, pricing is the area where
views are strongest and differences of opinion
widest. The pricing of natural gas has two aspects:
the price which BGC (and other potential purchasers)
are willing to offer producers (and exporters) -
generally referred to as a 'wellhead' or a 'landed'
price; and the price paid by the UK consumer
-domestic, commercial and industrial - for gas
supplies. The magnitude of both and the relationship that the consumer price should bear to the
landed price are matters of continuing, and at times
angry, debate.

Producer prices

Unlike oil, there is no 'world price' for natural
gas which can be identified. Also unlike oil,
natural gas prices (particularly in Western Europe)
are confidential: in the UK this applies to prices
that BGC is currently paying for both UKCS and
imported supplies. Of the three major world markets
for the fuel - Western Europe, the United States and
Japan - only the last could be said to have a
homogeneous pricing structure which applies to the
whole country. Elsewhere, the price of gas varies
greatly from country to country and state to state.
The prices of gas in individual international trades
are similarly varied, both within and between
markets. One of the most important issues in West
European gas trade in the 1980s is the extent to
which internationally traded natural gas prices
should bear a relationship to the international oil
price. (42)

The UK situation has been dominated by the position of BGC which, until recently, represented both a monopoly purchaser and sole vendor of natural gas. That situation was altered by the passage of the Oil and Gas Enterprise Bill (1982), but it is not yet clear whether, and how much, this may alter the actual extent of the BGC monopoly in both the purchasing and the sales functions. The central argument between the companies and BGC concerns the price which the latter has been willing to offer for supplies from the UKCS. The view of the companies has been that, throughout the 1970s, gas prices were too low to stimulate exploration and production.(43) In the 1980s, there is a recognition that BGC is willing to pay 'higher' prices and, although there is grumbling from the companies that the levels are still not high enough, there are indications that gas prices in newly negotiated contracts are in the range of 15-23 pence per therm. In September 1983 it was reported that BGC had been forced to pay 22-23 pence per therm for new North Sea gas in order to beat off a challenge from Imperial Chemical Industries which was attempting to secure the gas directly (as it is now entitled to do under the Oil and Gas Enterprise Bill).(44)

An important indicator of BGC's willingness to pay higher prices for imported gas came in 1980, when the Corporation offered a wellhead price for (Norwegian) Statfjord gas of $3.77 per million British thermal units (mmbtu), which would have translated to a landed price in excess of 25 pence per therm.(45) Despite the fact that Statfjord gas eventually went to the Continent, the BGC offer served as a reminder that prices of future gas imports would be considerably higher than those currently being paid (for Frigg gas), and exactly the same comment holds true for future gas supplies from the UKCS. This is illustrated in Table 6, which shows that the average price that BGC paid for gas and gas feedstocks had been rising faster in the late 1970s (with the introduction of Norwegian gas) than in the earlier part of the decade and had reached 11.64 pence per therm in 1982/83. The cost of prime materials in 1982/83 amounted to 36.2 per cent of the 32.17 pence per therm average cost (subtracting the gas levy this figure would rise to more than 40 per cent), compared with around 20 per cent in the early 1970s. From fragmentary data, it can be roughly calculated that the average price paid for UKCS gas in 1982/83 was 8.6 pence per therm which contains a large proportion of old contracts where the price was very low.(46) It is quite clear that new UKCS supplies from associated gas and condensate fields will cost considerably more to produce, compared with the relatively large dry gas fields of the Southern Basin. With UKCS gas accounting for at least 70 per cent of total supply up to 1990, the marginal cost of gas production will rise steeply and the average cost will follow, particularly in the last decade of the century as the Southern Basin fields go into rapid decline.

However, if the cost of UKCS gas is expected to rise sharply, the cost of the majority of available imported supplies is hardly likely to provide an attractive alternative. On a rough calculation, the price of Frigg gas had risen to around 22.5 pence per therm by the end of 1982, which gives a marker for the negotiations for Sleipner gas.(47) Sleipner may be considered the bridge between 'low-cost' Norwegian supplies (Frigg, Ekofisk and Statfjord) and high cost future supplies (Troll, Troms and other northern Norwegian fields).(48) Whether or not BGC does manage to secure Sleipner gas, there should be an indication of the price that the Corporation would be prepared to pay for base load supplies from Norway. The present author's guess would be that Norway would find it unacceptable to settle for a landed price of less than 30 pence per therm for Sleipner gas. Such a price raises the question of whether, if there are no customers at that price, Statoil (and the Norwegian Storting) will be prepared simply to delay development. However, Sleipner will be a landmark negotiation for both the willingness of BGC to pay higher gas prices and the attitude of Norway towards the timing of gas sales, the reason being that future Norwegian gas will be a great deal more expensive to produce.

Table 7 shows illustrative costs for new Norwegian gas. The total costs were not attributed to any particular field - but the author uses them in connection with a discussion on the development of Troll - and they do not appear to include any return on the gas. While these figures come from Statoil, and should therefore be considered in the light of a seller who is trying to prepare and per- suade purchasers into contemplating a very high price for future gas supplies, they cannot simply be dismissed as propaganda. It is generally accepted that the costs for developing Troll will be higher than for anything yet seen in the North Sea.(49)

In phase 1 of the development (we) envisage to install a single integrated platform with a year capacity of some 15 mtoe (about 17 BCM) of gas and 100,000 barrels per day of oil. The investment in this platform, including production wells and subsea facilities, will be in the order of $5 billion (1983).

The field could eventually require four platforms of this type.

Against the Statoil arguments it must be said that the volume of gas in the Troll field is extremely large and that, although initial investment costs may be high, the per unit cost should be considerably reduced by the huge volumes. Secondly, many people have suggested that the Troll field will require a major rethink on the part of the Norwegian Government in terms of taxation policy. Thirdly, Troll gas will not benefit Norway unless it can be produced at a cost, and hence offered at a price, which the purchaser can reasonably meet. The question of alternative Norwegian energy options without Troll gas is therefore extremely important.

Looking at other base load sources of supply, the question of the price of possible Soviet gas supplies is interesting. According to press reports, the price for Soviet gas delivered to the French border was agreed in early 1982 at $4.65 per mmbtu, which includes 30c per mmbtu transportation costs.(50) For illustrative purposes, one might take the French price and add a notional transportation cost for 600-700 km of pipe for the extension of the grid across France (and possibly Belgium), including a water crossing of up to 50 km in water depths of less than 200 metres, of 50-60c per mmbtu, bringing the c.i.f. price of Soviet gas on the UK mainland up to $5.15-5.25 per mmbtu (roughly equivalent to 32-33 pence per therm). It is possible that the USSR would be prepared to make concessions on the f.o.b. price to the UK in recognition of the higher transportation element.

The British bargaining position is made stronger since the UK is the only remaining large-scale

market in Western Europe in which Soviet gas plays
no part (assuming that the Netherlands will remain
self-sufficient for the foreseeable future). If the
agreement between the NATO countries on a limit for
Soviet gas of 30 per cent of total consumption of
any one country remains in force, it will be
difficult for the USSR to deliver additional
quantities to West Germany, France and Italy, prior
to the mid-1990s.(51) It would, of course, be
possible for it to sell gas to Sweden, Spain and
Belgium (the latter country possibly being the route
for the line to the UK), but the maximum sales
volumes would be limited, due to the small size of
the market in these countries. If, therefore, the
Soviet authorities wish to sell additional large
quantities of the fuel in Western Europe, their best
prospect will be to approach the UK with a large
quantity of gas at an attractive price. It is, of
course, easier for a non-market economy to
manipulate the price of an export commodity, and
while it is unlikely that the USSR will attempt
greatly to undercut the 'market price', it is
certain that Soviet gas will be priced very
competitively against all of its alternatives.

The question of Algerian gas pricing, either for
LNG or pipeline gas, is rather more complex. As
noted above, the UK imported Algerian gas for 17
years in what proved to be a satisfactory and
profitable trade for both parties. The contract
expired in 1980, but there was a nine-month
extension under which volumes were delivered at
$4.60 per mmbtu f.o.b. rising to $4.80 per mmbtu
f.o.b. in July 1981.(52) The parties were then
unable to agree a price for further deliveries and
the trade ceased at the end of 1981, except for one
spot cargo during the winter.

The price established for the nine-month
extension - equivalent to some 29-30 pence per therm
f.o.b. - was for a small volume of 0.75 BCM for a
short and limited period. In other words, these
could be regarded as limited peak load, as opposed
to (Norwegian) base load, supplies being delivered
through facilities where the capital cost had
already been amortised over the life of the contract. The inability of the parties to agree price
terms for a further extension was linked to Algerian
demands for parity with crude oil in the base price
and escalation clauses in any new contract. This

was a demand that BGC felt itself unable to meet, even for a peak load contract where the capital investment had been amortised.(53) Thus the prospects of BGC being prepared to sign base load contracts with Algeria, if the latter continues with its price claims, are not promising. With the success of Algeria in obtaining prices from Italy (pipeline gas) and France (LNG) which approach c.i.f. parity with crude oil and escalate directly with crude oil prices, there is no likelihood that it will be prepared to make a concession to the UK on gas pricing. Any indication that this was being considered in Algiers might be reflected in a resumption of the small-scale LNG trade through the Canvey Island facility.

High price demands, combined with previous instances where Algeria has demanded price renegotiation and curtailed gas deliveries in support of its price demands (in the case of the Continental contracts), would deter BGC from entering into base load supply contracts with the country in the immediate future.(54) However, it is worth repeating that, in the case of the original BGC contract, the experience over a long period of years was extremely favourable and that, given a change in Algerian Government policy and a number of years satisfactory trading experience with Continental countries at acceptable prices, there is no reason why Algeria sould not be reconsidered as a supplier of LNG or pipeline gas.

Prices for possible gas deliveries from the Netherlands would depend on whether peak or base load supplies were being contracted. However, it is likely that the UK could expect the same terms as Continental countries, with a small additional allowance for new pipeline links which would need to be constructed. While the cost of Dutch gas production and transportation will be low compared with other sources of supply, the Netherlands Government and Gasunie will expect the price to reflect the logistical and political attractiveness of Dutch imports. And, as noted above, the competition with Continental countries will be fierce.

All other basd load supplies would be more ex-pensive, in terms of their cost of production and transportation. Despite initial optimistic forecasts

33

of LNG availability and cost in the mid-1960s, when it was believed that 'imported gas could be landed on these shores from as far away as Nigeria and Venezuela for about 4½d per therm', LNG costs have escalated dramatically, in comparison with the cost of pipelining gas.(55) The huge cost of the Nigerian LNG Bonny project has caused the project to be considerably scaled down. The best option for the long term, ie beyond the turn of the century, appears to be Middle East gas, where very large quantities could be made available on a scale that might reduce the price per unit to an acceptable level, particularly if the facilities were being shared by Continental utilities.

Table 8, which summarises some illustrative costs and prices for domestic and imported gas, raises some essential questions for future supplies and the choices between available alternatives. First, it suggests that incremental imports from any source will be roughly three times as expensive as the current average cost of UK supplies and roughly four times the average price currently being paid for UKCS supplies. Second, it raises the question of how soon the cost of future UKCS supplies will rise to the level of imported supplies and whether UKCS will be preferred to imported supplies. Third, it suggests that UK prices will rise rapidly as the older, lower price contracts for UKCS gas expire, although not as rapidly as might be expected since, at present, the price of gas supplies only amounts to 40 per cent of the total price per therm charged to the consumer.

Consumer Prices

Consumer prices are the most directly politically sensitive area in the gas industry. Every government is aware that BGC's 16 million customers each have a vote (and some have more than one vote). Public opinion has proved to be extremely sensitive on the question of whether counterparts (or competitors) in other European countries are paying more or less for their gas than is the case in the UK (although this is usually very difficult to calculate and changes rapidly over time due to exchange rate fluctuations). There is a sizeable slice of public opinion which holds that it is right and proper for UK citizens to pay less for 'our gas' than less fortunate Continental countries which have to import supplies. Such a view takes little cognisance of the effects of the price mechanism on conservation, or the rising cost of future supplies. Emphasis is often placed on the problems and privations suffered by poorer sections of the community which are disproportionately affected by rising gas prices. This is an extremely important social policy question, which should be the primary responsibility of the Department of Health and Social Security rather than the Department of Energy, although BGC has become increasingly involved in the problem in recent years.(56)

The two major features of consumer gas prices are, first, that they have generally been cheaper (particularly in the residential market) than competing fuels, and secondly that they have repeatedly been manipulated by successive governments for political and economic (mainly revenue-raising) reasons, which have little to do with energy policy. To illustrate this point with some recent examples:(57)

a) constraint of prices which mainly affected domestic prices during the period when the Price Commission operated (1973-9);

b) constraint of domestic prices in the autumn of 1979;

c) the sequence of three domestic tariff increases of 10 per cent per annum completed in 1982;

d) the freezes on industrial and commercial prices during the period since 1980/81.

Under these influences, average revenue per therm in real terms behaved as follows in the period 1975/76 to 1981/82;

a) domestic tariff fell by 20 per cent, then recovered most of the fall;

b) non-domestic tariff approximately retained its level;

c) commercial firm contract increased by 60 per cent;

d) industrial firm contract increased by 100 per cent;

Over the same period the cost of gas increased in real terms by 200 per cent (150 per cent excluding gas levy), and gross profit per therm declined in real terms by almost 20 per cent. It is very difficult to perceive any coherent

marketing or commercial logic in this pattern of
cost and price movements.

There appear to be two major schools of thought
on the subject of consumer pricing. One holds that
natural gas prices should be set by the market, with
supply and demand determining the level of prices
and gas competing with other fuels in the market-
place. The other is that the price of gas to the
consumer should, under present policy, bear a
relationship to the expected long-run marginal
landed price. Reference to the long-run marginal
landed price is not so much an analytical economic
approach, as an attempt to identify the cost of the
replacement therm, whether imported or domestically
produced.

These concepts are introduced here tentatively
because they do not appear to bear any very close
relationship to reality. First, there is a very
imperfect market place for energy; the majority of
industrial and (particularly) domestic consumers
cannot switch from one fuel to another with
impunity. Secondly, if the price of natural gas had
risen to match that of oil during the two price
shocks of 1973/74 and 1978/79, the domestic hardship
would have been enormous. The marginal concept is
equally difficult to interpret: Deloitte Haskins and
Sells suggest, in their report on the activities of
BGC, that consumer prices should reflect marginal
costs which they define as the cost of Norwegian
Frigg gas plus a 5 per cent real rate of return.
Yet there is no reason to select Norwegian Frigg gas
rather than Sleipner gas or SNG from coal as the
marginal source, or 5 per cent as the correct rate
of return.(58) Consumer prices based on the cost of
any one source of gas which is only a small part of
total gas supply - 20-25 per cent in the case of
Norwegian Frigg - will be grossly distorting and
inevitably involve major discontinuities when the
'replacement' marginal source has to be chosen.

Recent gas pricing policy appears to have aimed
at a compromise between these approaches (although,
as outlined above, actual policy has been subject to
short-term political and economic imperatives). The
aim has been to raise the price of gas to a level
which reflects both marginal cost, defined very
broadly with a large number of inputs averaged in,
and the market price, defined so that the price of

36

gas should not be greatly out of line with that of crude oil. Furthermore, there has been an attempt to raise the price in small steps, rather than large leaps, so as to minimise both hardship to consumers and domestic political reaction.

However, the most important departure raised by the emphasis on gas pricing is the view of the Deloitte report that 'if the objectives of BGC, notably pricing policy, are clearly defined, a separate depletion policy is unnecessary'.(59) A major recommendation of the report is that a long-term pricing policy should be agreed between BGC and the government, despite the cataloguing of deliberate government intervention in the industry for mainly political reasons.(60) For those who believe that price is the most important factor in natural gas policy, self-sufficiency is a side issue and any kind of controlled depletion policy is irrelevant. This would be a distinct change from the policy which BGC has adopted up to the present, which places considerable limitations on gas sold to non-premium markets and refuses to countenance exports. If government policy were to be changed to allow exports of UK gas to go ahead, this would undoubtedly affect consumer pricing in the short term.

Finally, it is worth stressing three long-running obstacles to any serious discussion of gas pricing policy. First, the problem that natural gas contract prices, both in the North Sea and Western Europe in general, are shrouded in confidentiality; it is high time that a greater degree of transparency was accepted by all parties.(61) Secondly, there must be doubt that future costs of domestic or imported gas can be projected with any degree of confidence. Thirdly, and probably most important, there seems no hope of achieving consistency in future consumer pricing policy; there is every indication that governments will continue to pursue short-term political and revenue-raising objectives, in preference to energy/gas pricing considerations.

The Demand Situation

The demand scenario which has been chosen for the BIJEPP self-sufficiency project is scenario 'BL' from the Department of Energy's <u>Proof of Evidence for the Sizewell 'B' Public Inquiry</u>. This is the base case, which will be used here in the knowledge that a single scenario is bound to be artificial, but that for illustrative purposes it is useful to have a general reference point from which to start. The scenario sees natural gas demand rising marginally from around the 1982 level of 47 BCM to nearly 50 BCM in 1990, falling back to 47 BCM by 2000 and back further to 44 BCM by 2010 (Table 9). If one extends this trend, it suggests a further fall to 40 BCM by 2020. This particular scenario is predicated on three assumptions: an oil price of $43 per barrel by 2000 (which the Department terms 'low energy prices'), 1.5 per cent per annum growth of GDP, and low industrial growth. Aside from the general hazards of demand forecasts, the Department notes specifically:(62)

> The analytical approach used, which is dependent in large part upon historical experience, is unable to reflect fully the effect of opening up this market to greater competition (as a result of the Oil and Gas Enterprise Act, 1982) and, in particular, the removal as a result of the Act, of current unsatisfied demand.

The gas price scenario (Table 10) which corresponds with the demand forecast outlined above indicates that the fuel will remain competitive in the domestic market and that it is in this sector that natural gas supplies will remain overwhelmingly concentrated. From this scenario it appears that the fuel has a reasonable prospect of holding its share of the commercial and public service market, although here, and more particularly in the industrial sector, the price disadvantage with coal will be considerable.

The scenario suggests a continuation and accentuation of UK gas consumption policy since the arrival of North Sea gas in the late 1960s. Unlike a number of Continental countries, where gas has had a sizeable role in power generation and a significant share of the industrial market, the UK has given priority to premium markets (which in practice has meant the domestic sector and a small section of commercial/industrial requirements) and confined sales outside the premium markets to the minimum necessary to make up the load (Table 11).(63) It is clear that, in the domestic sector, gas is best equipped to withstand competition from other fuels and the Department of Energy sees consumers continuing to move to gas from other fuels for space and water heating purposes.(64) The Department suggests that the use of gas for non-energy purposes, primarily petrochemical feedstock, will decline steadily through the period to zero by 2010.(65) Some further observations will be made on demand levels below, but it is important to consider UK natural gas self-sufficiency in the context of this demand level.

UK Natural Gas Supply Options to 2020

This section will suggest the range of options for
UK gas supplies to 2020, trying to identify the
major decisions, the timing of those decisions, and
their implications for self-sufficiency. The method
will be to consider two options for UKCS production
and then to consider the additional requirements
which stem from these production levels. The two
variants are as follows:

   i)   High UK Continental Shelf supply. In this
        variant, UKCS production rises slowly to 40
        BCM per year in 1990, 50 BCM in 2000,
        falling back to 40 BCM by 2010 and 30 BCM
        by 2020. This variant exhausts the highest
        known level of UKCS recoverable reserves
        (as defined in the 1983 Brown Book) by
        2020.

  ii)   Low UK Continental Shelf supply. In this
        variant, UKCS production remains at 35 BCM
        up to 1990 and then falls steadily to 28
        BCM in 2000, 20 BCM in 2010 and 15 BCM by
        2020. At this level of production, some
        500 BCM of recoverable reserves (as defined
        in the 1983 Brown Book) remain to be
        produced.

It is not suggested that either variant reflects a
likely outcome. Rather, the two variants are
suggested parameters, within which the actual
situation may fall. Taking the Department of Energy
demand scenario outlined above, the picture would be
as follows:

	High UKCS Supply		Low UKCS Supply		Demand (a.)
Year	% Supply from UKCS	Require- ment from other sources BCM	% Supply from UKCS	Require- ment from other sources BCM	BCM
1990	80	10	70	15	50
2000	100	–	57	19	47
2010	91	4	50	22	44
2020	75	10	38	25	40

(a.)   Department of Energy 'BL" scenario except figure for 2020 which is author's extrapolation.

High UKCS Supply/High Priority Self-Sufficiency

It is clear that a high UKCS production would be followed as part of a policy where the prime objective was the maintenance of maximum self-sufficiency for the longest possible time. UKCS production would rise substantially during this period and be maintained near the current level up to 2020. Most people would believe this to be a very optimistic scenario in the absence of significant additional dry gas being discovered. However, although it would be a theoretical possibility, two points should be recognised. First, after 2000 when the large dry gas fields in the Southern Basin are in decline, production totals would rely heavily on small associated gas and condensate fields. Secondly, under this scenario, natural gas production after 2020 would fall to zero immediately. Therefore, while the high variant prolongs a comfortable level of self-sufficiency up to 2020, it would give rise to a very sharp discontinuity immediately thereafter, when the country would be forced to rely almost totally upon other sources of supply.

In addition, if the promotion of the greatest possible degree of self-sufficiency up to 2020 (without regard for the position thereafter) were the prime policy objective, the aim would be to promote high UKCS supply together with the fastest possible development of SNG from coal. Given the lead times of 10-15 years for the building of coal gasification plants and the current stage of development of the technology, it seems inevitable, even in the high priority self-sufficiency case, that some additional imports would be required for at least the early part of the final decade of this century. Thereafter UKCS supplies would satisfy demand until the end of the first decade of the next century when two SNG plants (each of 2.5 BCM per year capacity) would be required. By 2020, there would be a requirement for four SNG plants and subsequently a much larger number to cope with the imminent fall in UKCS production. Nevertheless, this variant sees the UK virtually self-sufficient in gas supplies from 2000 to 2020. The high UKCS variant figures in the table above assume that the UK does not contract for further Norwegian (Sleipner) gas supplies. The likely volumes of 9-11 BCM per year of Sleipner gas, starting in the early 1990s, would eliminate the need for any further supplies supplementary to UKCS production, until well into the second decade of the next century.

High priority self-sufficiency is likely to be the highest cost gas supply option, with SNG production costing more than 50 pence per therm (in 1980 money), which is likely greatly to exceed the cost of imports from most sources at least up to 2000 and probably beyond. High costs would be translated into high prices, which would inevitably mean a reduction in demand levels. Thus high priority gas self-sufficiency is likely to result in a contraction of the gas market in the long term. Another consequence of high priority self-sufficiency is the situation after 2020. As mentioned above, on present estimates such a level of production would exhaust known recoverable reserves by 2020, after which the industry would have to make a rapid transition to total reliance on SNG from coal. The alternatives would be to abandon self-sufficiency as a policy (and switch to imports) or abandon gas as a major fuel. Total reliance on SNG after 2020 would suggest a need for 15 coal gasification plants with a total requirement of 75 mt of coal per year. The scale and rapidity of the transition from UKCS supply to SNG would suggest a major break in the logistics of the gas industry at that time.

Low UKCS Supply/Low Priority Self-Sufficiency

Low UKCS supply would be allowed under a policy where self-sufficiency was not considered to be a priority. Supplies would be drawn from available sources, largely on the basis of price with attention to possible over-dependence on any one external source of supply. Under low UKCS supply, additional gas requirements would rise steadily from 15 BCM in 1990 to 25 BCM by 2020, but might remain steady or fall thereafter as production could be maintained. Self-sufficiency would decline rapidly from around 70 per cent in 1990 to around 50 per cent by 2010 and less than 40 per cent ten years later. The balance of supplies could be made up of various permutations of pipeline gas from Norway, the Netherlands, the USSR and the Middle East (from Egypt to Qatar) and/or LNG from a number of sources including Middle East and West African countries plus the Canadian Arctic, Trinidad and South America.

Looking at the availability of gas from different suppliers, it is likely that there will be a wide choice of sources open to the UK, to the point where careful selection becomes necessary. Around 2020, one can easily imagine the UK having access to a choice of: 20 BCM of Norwegian gas, 20 BCM of Soviet gas, 10 BCM of LNG supply from various countries and 10 BCM of Middle East gas, making a total of 60 BCM, which is 50 per cent greater than the projected total demand level in that year. If the UK does turn out to be in the happy position of being able to choose between different sources of supply, the price element may be decisive. From Table 8 the indication is that SNG from coal will find it difficult to compete with any of the sources of supply illustrated in the table, other than the most expensive new Norwegian gas.

All other sources appear to be broadly competitive with each other, except for Dutch gas supplies and Middle East pipeline and long-haul LNG imports on which it is difficult to put even an illustrative value.

To sum up the two cases. In the high variant, Norway would be the only external source of gas required up to 2020, but imports would cease in the early 1990s (after the expiry of the Frigg contract) and the first SNG plant would be on stream by 2010. In the low variant, Norway would be the main (and perhaps the only) external source, aside from marginal LNG imports, through the mid-1990s, but a decision would need to be taken around the beginning of that decade, on how demand in the late 1990s and the early part of the next century was to be met. The choice appears to be between additional imports from Norway, supply from the USSR via a cross-Channel pipeline, and LNG supplies from Algeria (among others) via a new UK terminal.

In both variants, Norway has been used as the cornerstone of non-UKCS gas supplies. This judgement is based on BGC import policy, the abundance of gas which Norwegian reserves indicate will be available from the country over the next several decades, and the close political relationship with an OECD and NATO ally, which contrasts sharply with any other source of potential gas supply (the Netherlands excepted). The potential problem with this course of action may be the high cost of future Norwegian gas production, which may disadvantage this source in comparison with Soviet gas. Of the possible external sources, only Soviet gas is likely to enjoy a <u>considerable</u> price advantage over Norwegian gas.

However, it is not certain that Soviet gas would be politically acceptable, despite the prospect of its being offered at competitive prices. The future of the trans-Atlantic debate over the political wisdom of importing gas from the USSR may be an important factor in determining whether this source of energy is considered seriously in the UK. Certainly, up to the present time, this has not been the case. Soviet gas was not considered as an alternative to Norwegian Sleipner gas in the early 1990s, although it is certain that a comparable quantity of gas could have been available and the Soviet authorities would have been amenable if approached. A BGC official admitted:(66)

I have to say that at present we have not directed our minds very closely to the idea of importing gas from Russia. As years go by, maybe that option will become more prominent in our thinking. However, I have to say that at the moment it is not prominent in our thinking. We tend to look firstly towards Norway and secondly - and, I think, less desirably - to LNG imports from further afield.

In general, neither the Department of Energy nor BGC can bring themselves to mention the USSR by name in official publications, and at least part of the reason for this is the political unpopularity of such an idea with the present Conservative Government.(67) It is clear that considerable political opposition would have to be surmounted before Soviet gas would be considered a real option for the UK. If it should continue to be excluded from the UK market for political reasons, this would open the way for higher cost LNG supplies from a number of sources, and/or increased high-price Norwegian supplies. Without Soviet gas supplies, a cross-Channel pipeline seems unlikely until the second decade of the next century when pipeline supplies from the Middle East may be a real prospect. However, under the low UKCS supply scenario with no restrictions on sources of supply, it is likely that the additional 25 BCM required by 2020 would be supplied by Norway (15 BCM) and the USSR (10 BCM).

Questioning Assumptions: Some Reflections on Reserve, Demand and Price Scenarios

Given the uncertainty of reserve estimates, the artificiality of taking one energy demand and price estimate, and the likely inaccuracy of all scenarios, even in the short term, let alone looking forward four decades, it is important to reflect on the possibility of some radically different outcomes and the effect of these upon UK self-sufficiency.

Most obviously, if the high reserve figures in Table 1 prove to be correct, it should be possible to maintain UKCS production at a high level -perhaps as high as the high supply case outlined above - for the duration of the period up to 2020, without suffering a catastrophic fall thereafter. However, it is important to consider whether these additional reserves are likely to be discovered in large dry gas accumulations - akin to the Southern Basin fields - or whether they are more likely to be in smaller fields; the geographical location and water depth in each case are also likely to be important. All these factors will have a considerable bearing on costs of production and hence the competitiveness of UKCS gas against external sources of supply. The Brown Book estimate of reserves is conservative, but if production is raised on the assumption that greater reserves exist than are actually located in the future, such a policy may have serious implications for self-sufficiency. If, however, larger reserves than expected are located, then this will be a pleasant surprise which will take the pressure off the selection of UK gas supply options in the future.

On the demand side, the present author's intuitive guess would be for a higher gas demand than in the scenario described above. A different scenario is shown below under which demand rises from 50 BCM in 1990 to 60 BCM by 2000, falling to 55 BCM by 2010 and to 50 BCM in 2020.

Higher Demand Case (BCM)

	High UKCS production	Additional requirement	Low UKCS production	Additional requirement	D
1990	40	10	35	15	50
2000	50	10	28	32	60
2010	40	15	20	35	55
2020	30	20	15	35	50

D   Demand levels which are somewhat higher than the Department of Energy's highest case in their Proof of Evidence to the Sizewell Inquiry. That case was based upon an oil price of $52 barrel in 2000, 2.5 per cent growth per annum and high industrial growth.

Price will continue to be a critical issue since, in order to make a high demand case credible, one must posit a gas price which allows the fuel to compete successfully in end-use markets. Up to the present time, the large proportion of UKCS gas in total supplies has meant that price could be held down in order to encourage demand penetration. Even the rapid rise in prices, as a result of government policy, over the past three years has not dramatically affected gas demand, which has held up well in a time of falling energy demand, in comparison to other fuels (Table 12). In the future, particularly by the 1990s and beyond, the cost of natural gas from domestic sources - whether UKCS or SNG from coal - will rise considerably. The most likely source of lower cost gas supplies, therefore, is from imports, and the growth of gas demand after 1990 may be importantly linked to the availability of attractively priced pipeline imports.

In addition to the price issue, the additional requirements of the higher demand case suggest a revision of the timing of decisions on supplementary gas supplies. The figures in the higher demand case above make the high self-sufficiency scenario rather more complicated, although by no means impossible. In order to maintain total self-sufficiency, four SNG plants have to be completed by the end of the century, six by 2010 and eight by 2020. In contrast, the scenario below, showing possible choices between supplementary gas supplies under a high demand scenario, abandons self-sufficiency as a priority and concentrates on selection and timing of imports.

Choices Between Large-Scale Supplementary Gas
Supplies: Higher Demand Case(a)

Date of Choice	Date of Require-ment	Quantity (BCM/ Year)	Choice
1983/4	early 1990s	10-15	Norwegian Sleipner gas or increased UKCS gas; possible small-scale Dutch supplies
early/ mid- 1990s	2000	10-22	Depends on progress of Norwegian Troll Field and likely price of Troll gas. Quantity is so large that some gas likely to come from Troll plus around 10 BCM from USSR. If either of these is unavailable or unacceptable, large-scale LNG option will need to be considered.
late 1990s	2005- 2010	13	Increased Norwegian/Soviet/LNG supplies depending on previous choices. Possibility of commencing large-scale SNG production.

(a)Assumptions: all projects require 5-10 year lead time from contract signing, all contracts last for 20 years.

This scenario therefore reflects a low self-sufficiency priority (with SNG being considered as a possible option only towards the end of the first decade of the next century) and an acceptance of large-scale imports.

The major difference between the import requirements in the higher demand case and those in the base case scenario is that the higher figures require the elaboration of a more careful import strategy. If the UK is importing 40-50 per cent of its natural gas supplies at the end of the century and as much as 70 per cent of supplies by 2020, then diversification of suppliers will be an important part of the strategy. While it may be highly satisfactory that Norway supplies all UK external requirements, through the mid-1990s, it is questionable whether any one country should be providing up to one half of total gas demand at the end of the century. Indeed, there is a good case for sticking to the trans-Atlantic agreement formed out of the dispute over Soviet gas supplies, that West European countries should avoid 'undue dependence upon' any one external gas supplier. In practice, this means a limit of 30 per cent on imports from one source, and is an important security measure to eliminate vulnerability caused by a sudden cessation of supplies from a single source. The UK would be particularly vulnerable to such a cut-off in the winter months when average daily demand can be as much as three times the summer figure. Therefore, if the UK should be looking for an additional 35 BCM of imported supplies in the context of a total demand of 50 BCM by 2020, 60 per cent of supplies could be provided by Norway and the USSR, and, with an additional 30 per cent from the UKCS, this would leave 5 BCM to be found from elsewhere. This would require the addition of an LNG and/or SNG option, which would allow further diversification if necessary. The actual proportions provided by each source would be decided on price competitiveness.

As mentioned earlier, the policy up to now has been to restrict natural gas consumption to largely domestic and high value commercial and industrial uses, with sufficient additional industrial (interruptible) demand to balance the load. It might be argued that for natural gas demand to rise in the manner of the high demand forecast above, the fuel would need to penetrate the domestic market further (which would seem eminently possible) and at least hold its own in the commercial and industrial sectors, which means being able to compete with electricity, gasoil, low sulphur fuel oil and, in some sectors, coal. The critical areas of competition may be electricity in the residential market and coal in the industrial market. For industrial users, coal may become a better alternative than oil as a substitute for gas, depending on the relationship between gas and oil prices. This situation will be considerably affected by the progress towards rationalisation in the UK coal industry and the possible effect on the price of coal of the current concern over excess sulphur in the atmosphere and the environemental effects of acid rain.

If the price of gas should make continued competition in these sectors impossible, then one might expect a contraction, or at least no great increase, in gas demand.(68) This might give rise to a different gas usage strategy, under which all industrial contracts would be eliminated and the market would be restricted to the sector which was best equipped to withstand high prices, ie residential demand. The problem with such a strategy is that it might not be technically possible for BGC to handle the peak load periods without the facility of interruptible customers. It would require, at a minimum, that storage capacity be dramatically increased (which would further raise the per unit cost to the consumer) and that the load factor on supplementary supplies be very flexible.

In the domestic residential sector, one would expect natural gas demand to come under pressure from conservation measures, particularly as the recent gas price rises work their way through to consumer behaviour. It is interesting that in the case of the Federal Republic of Germany:(69)

> more than half the energy saving by residential gas users is not attributable to new or improved equipment or facilities such as double glazing, but to changes in consumer behaviour, such as a reduction of indoor temperature levels or the elimination of energy wastage by not heating unused rooms.

In addition, the Economic Commission for Europe sees a considerable potential role for gas heat pumps in Western Europe prior to the year 2000.(70) It is noteworthy that the lowest of the Department of Energy scenarios gives a demand level of only just over 30 BCM in 2010 which, even in the low UKCS production case suggested here, would suggest that imports in that year would not need to be above current levels.

A Note on Northern Ireland

Before summing up the UK position on self-sufficiency and the options for the future, it is worth recalling that there is a part of the UK which is facing total dependence on imported gas. Northern Ireland's small town-gas industry, composed of thirteen undertakings, is largely dependent upon hydrocarbon feedstocks (naphtha and LPG) and became uncompetitive after the first oil price rise of 1973/74.(71) In 1976, BGC carried out a study which confirmed the uncompetitive position of the industry and examined a number of alternatives: piped gas from Scotland, piped gas from Eire, importation of LNG and importation of LPG. The option which received most discussion and comment was a pipeline from Scotland which would have brought North Sea gas to Northern Ireland. The BGC report concluded that a link to Northern Ireland would be uprofitable, resulting in losses amounting to £87.5m. by 1989. As a result of this finding, the government announced, in mid-1979, that pipeline gas to Northern Ieland could not be justified, nor could the continued subsidisation of the industry, which would be assisted to run down in an orderly fashion.

A number of organisations in the province took issue with the BGC conclusion and commissioned their own report, which suggested a more favourable economic climate for piped gas supplies.(72) The view of one group, committed to maintaining employment in the industry in Northern Ireland, was as follows:(73)

It is notable that other areas of the UK more remote than Northern Ireland have received supplies of natural gas as of right.

Discrimination against Northern Ireland in this matter is scandalous.
(In fact, BGC has no obligation to supply gas 'as of right' and there are many areas of the UK which do not have a piped gas supply.)

The conclusion of a careful and well-balanced study of the question was that pipeline gas from Scotland to Northern Ireland could be economic, but that a decision should be made dependent upon negotiations with the Republic of Ireland on the cost and availability of gas supplies from the Kinsale field, which are being piped to Dublin via a line which could be extended north to Belfast.(74) The projected gas demand was estimated at around 80 million cubic metres in the early 1980s, rising to 275 million cubic metres by 1989. These are not large figures and would not have constituted a great drain on UK reserves, had the pipeline from Scotland been built.

In October 1983, agreement was reached on gas sales to Northern Ireland from the Kinsale field, via a pipeline to be constructed and operating by 1985; the contract would run for 22 years.(75) Despite the claim by the Irish Minister for Industry and Energy that 'no political overtones have entered the discussion. The agreement was negotiated on commercial considerations', it seems clear that there was a considerable political incentive on the British side to achieve an agreement, both in terms of improving relations with the Dublin Government (and creating long-term links between Northern and Southern Ireland) and in creating employment in Ulster.(76) The commercial considerations referred to by the Minister appear the shakiest part of the agreement, with the issue of the price of the gas not yet settled (although a figure of 26 pence per therm has been mentioned). The agreement is likely to go some way towards assuaging the province's long-standing grievance that it has been deprived of the benefits of North Sea oil and gas. However, it is interesting that an opportunity to extend natural gas self-sufficiency to Northern Ireland was rejected and that the British Government has encouraged the province to become totally dependent upon an outside source for its gas supplies, which will eventually constitute some 12 per cent of total energy supplies, through a facility which may be vulnerable to terrorist attack.(77)

The UK currently produces around 75 per cent of its
natural gas supplies and this level of self-
sufficiency will remain roughly constant through the
early 1990s. After 1995, and particularly after
2000, the picture inevitably becomes less clear. On
present knowledge of reserves, the Southern Basin
fields will begin to decline rapidly by the
mid-1990s and it will be necessary to replace this
production by exploiting a large number of associated gas and condensate deposits. These new
fields will require a much more complex geological,
engineering and logistical development, with longer
lead times and larger (per unit) investments, than
the dry gas fields which they replace. At a guess,
new fields would require a price in excess of 25
pence per therm - compared with a current average
price of around 8.5 pence per therm - indicating a
threefold real price rise for indigenously produced
gas by the end of the century.

At the highest level of reserves currently
estimated by the Department of Energy, it would be
theoretically possible for UKCS production to rise
from the current level of 38 BCM to 50 BCM in the
year 2000, slipping back to 30 BCM by 2020. This
production level would exhaust the highest level of
estimated recoverable UKCS reserves by 2020.
However, it would mean that the country would be
virtually self-sufficient in natural gas during the
period 1990-2010 and could extend that self-
sufficiency to 2020 by the construction of four SNG
plants using coal as feedstock. This total self-
sufficiency strategy is likely to be very high cost,
with SNG production estimated at more than
50 pence per therm - a price currently well above

projected prices of domestically produced gas and/or imports from any source. After 2020, SNG would have to take over gas supplies almost entirely, which, with a demand of 40 BCM per year, would require 16 SNG plants with an annual coal consumption of 80 mt.

An alternative low UKCS production path might see a level of 30-35 BCM in 1990, falling to 15 BCM by 2020, which would leave sufficient reserves remaining for production to continue at this level, perhaps up to the middle of the next century. Under this scenario, self-sufficiency would fall to less than 60 per cent by 2000, 50 per cent by 2010 and less than 40 per cent by 2020. The resulting gap in supplies would be filled by imports in the short to medium term; SNG would not be considered a large-scale option until the second decade of the next century. The low self-sufficiency scenario is more attractive on economic grounds, since the major available sources of imported gas are likely to be less expensive than SNG. Slower depletion of UKCS reserves indicates a gradual shift from indigenous to imported and/or SNG supplies, rather than a sudden transition.

The low self-sufficiency case would require the careful elaboration of a natural gas policy which balances domestic supplies against imports and weighs the various sources of imports against each other. The attraction of a continued high level of self-sufficiency is that it minimises vulnerability to the political, economic and technical vagaries of the outside world. That is certainly the present situation, with the UK having the capacity to withstand a protracted interruption in Norwegian supplies by utilising storage capacity and increasing production from domestic fields.(78) With increasing dependence on external sources, the capability to withstand such interruptions diminishes and hence a more careful strategy is needed. UK natural gas import policy has thus far considered Norway as the sole large-scale source of supply outside the UKCS. While this has been entirely understandable and realistic, given the fields which straddle the median line in the North Sea and the close economic and political relationship with Norway, it will need to be rethought for the period after 1990.

In the future, there will need to be a thorough evaluation of supplies from the USSR, which could be landed in the UK in large quantities at an attractive price. The only logistical addition required would be a pipeline across the English Channel, a link thus far resisted by BGC because of fears that it could be used to export UKCS gas. On present estimates of reserves and demand, large-scale exports would gravely endanger the supply position and bring forward the time when the country becomes a large-scale importer. Nevertheless, a cross-Channel pipeline which was designed for large-scale imports could also have a capability to provide emergency gas supplies to Continental European countries in times of crisis.

Norwegian and Soviet gas supplies are considerably more attractive in the short to medium term than large-scale gas supplies from any other source - pipeline or LNG. At present, Algerian LNG pricing policy does not make it attractive even to renew peak load supply which ceased in 1981. A new, large-scale, LNG terminal would require at least a decade to construct and while it would be possible for it to receive gas from a number of sources - Arctic Canada, Nigeria, Cameroon, Trinidad, etc - these are all long-haul (mostly small-scale) LNG supplies which would represent very expensive gas landed in the UK.

After the turn of the century, and certainly after 2010, the question of large-scale gas imports from the Middle East becomes a possibility and so do large-scale coal gasification projects. However, both of these sources are likely to be more expensive than any other large-scale pipeline supplies thus far mentioned and their viability will be determined by levels of demand and price. In the base case demand scenario, the 25 BCM of additional gas needed by 2020 could be covered by just two suppliers: Norway and the USSR. A lower demand scenario - caused by high gas prices relative to other fuels, and/or successful energy conservation in the domestic sector - could leave the industry with an external requirement in 2020 no greater than at present. A higher demand scenario, giving an additional requirement from non-UKCS sources of 35 BCM, opens up the possibility of three or four different sources of supply, including a large-scale LNG option, pipeline gas from the Middle East, and large-scale coal gasification. If any of these options were to be ruled out, ie Soviet gas for political reasons, LNG for environmental reasons, SNG on technical or cost grounds, then supply options become more restricted. Vital in all of these considerations is an appreciation of the lead times: (Norwegian) Sleipner gas being contracted in 1984 probably could not start flowing before 1992; a cross-Channel pipeline may require a similar time lapse; large-scale LNG and SNG options will require at least ten years from initial proposal to start of operations.

In summary, the argument in this paper would conclude that, even if it is possible to produce natural gas from the UKCS at the highest levels indicated above - allowing total self-sufficiency up to 2020 with the addition of a small quantity of SNG from coal - such an option should be resisted. In addition to being the highest cost strategy open to the industry, it runs the risk of severe supply discontinuity after 2020 which could, at worst, eliminate natural gas from the UK energy balance after that time. The preference would be for a lower UKCS production profile, along the lines of the low UKCS production variant outlined, with a gradual slide into greater dependence on imported sources, which could be halted at around 40 per cent self-sufficiency in 2020, a level which could be maintained up to the middle of the century. The volume of available reserves and production can be reviewed at regular intervals. If sufficient reserves are established to maintain a higher level of UKCS production than the low production variant, followed by a gradual decline, a higher level of self-sufficiency can be maintained for a longer period of time. Every effort must be made to avoid a sharp discontinuity in sources of supply. While greater dependence implies increased vulnerability, there are a large number of supply options from which choices can be made, contingent upon a mixture of security of supply and price considerations. So long as the length of lead times for individual options is appreciated, there is no reason to believe that gas will become a serious area of vulnerability in the nation's energy balance up to 2020.

NOTES

(1) Trevor I, Williams, _A History of the British Gas Industry_. Oxford University Press, 1981, p.296.

(2) Since the early 1970s the Department of Energy has produced a yearly publication entitled _Development of the Oil and Gas Resources of the United Kingdom_, colloquially known as the 'Brown Book'. It will be so designated in this paper with the appropiate year.

(3) Ray Dafter, 'Jobs Boost from Morecambe Field', _Financial Times_, special on The North West, p.V.

(4) International Energy Agency, _Natural Gas: Prospects to 2000_. Paris: IEA/OECD, 1982, p.69.

(5) Robert J Enright, 'Saudi Desert Sprouts Gas Plants', _Oil and Gas Journal_, 9 July 1979, pp.63-9.

(6) Martin Lovegrove, _Lovegrove's Guide to Britain's North Sea Oil and Gas_ Cambridge: Energy Publications, 2nd Edition, 1983, p.103.

(7) Department of Energy, _Gas Gathering Pipeline Systems in the North Sea_, Energy Paper No. 30. London: HMSO, May 1978.

(8) Department of Energy, _A North Sea Gas Gathering System_, Energy Paper No. 44. London: HMSO, June 1980; also, W. J. Walters, R. H. Wilmott, I. J. Hartill, 'The Northern North Sea Gas Gathering System', _Institution of Gas Engineers_, Communication 1149 , 1981.

(9) Sue Cameron and Ray Dafter, 'North Sea Gas Pipeline Plans Changed', _Financial Times_, 4 April 1981.

(10) 'Gas Gathering Pipeline', Department of Energy Press Notice, Reference No. 152, 11 September 1981; Ray Dafter, 'Anatomy of a 2.7bn Decision', _Financial Times_, 1 September 1981.

(11) 'UK Statfjord Gas to be Piped Ashore Via FLAGS', Lloyds List, 17 January 1983.

(12) 'Shell unveils the route for the Fulmar line', World Gas Report, 29 August 1983.

(13) A North Sea Gas Gathering System op.cit., pp.26-32.

(14) C. H. Bayly and T. F. Cox, The Economics and Politics of Gas/Condensate Gathering in the UK North Sea. Byfleet: Gaffney Cline and Associates, 1983.

(15) North Sea Oil Depletion Policy, Third Report from the Select Committee on Energy, Session 1981-82. London: HMSO, 7 May 1982, and North Sea Oil Depletion Policy: The Government's Observations on the Committee's Third Report of Session 1981-82, Sessions 1982-83. London: HMSO, 21 December 1982, pp.x-xi.

(16) This does not include associated gas used on production platforms (see Table 4).

(17) The Background to the Far North Liquids and Associated Gas Systems, Shell Press Release, May 1982.

(18) Calculated using Brown Book, 1983, Table 8, p.22 and subtracting Algerian imports.

(19) Norwegian Royal Ministry of Petroleum and Energy, The Norwegian Continental Shelf, Fact Sheet, 1983:1, pp.27-28.

(20) The primary difficulty is one of a shallow reservoir spread over a very wide area, suggesting very difficult drilling conditions. C. E. Fay, Development Prospects in 350 Metres Water Depth in Block 31/2, An Address to the 5th Offshore Northern Seas Conference and Exhibition, Oslo, August 1982, p.3.

(21) Norwegian Petroleum Directorate, Petroleum Outlook. Stavanger, 1982, pp.42-5.

(22) Peter Hinde, Fortune in the North Sea. London: G. T. Foulis & Co, 1966, p.169.

(23) Considerable speculation still surrounds this prospect but the beginning of a change in policy can be found in, N. V. Nederlandse Gasunie, 1983 Gas Marketing Plan.

(24) At 1 January 1983, Cedigaz estimated Soviet gas reserves at 34.5 trillion cubic metres, 39.8 per cent of world proven reserves. Cedigaz, Le Gaz Naturel Dans Le Monde, en 1982, Paris: Cedigax, June 1983.

(25) There would appear to be around 10 BCM of spare capacity in the Soviet pipeline network to Western Europe in the early 1990s, unless additional

contracts are signed with Continental countries.

(26) Further details can be found in Jonathan P. Stern, _International Gas Trade in Europe: The Policies of Exporting and Importing Countries_, BIJEPPP Energy Paper No. 8. London: Heinemann for RIIA/PSI, 1984, Chapter 7.

(27) Department of Energy, _Energy Policy: A Consultative Document_, Cmnd 7101. London: HMSO, 1978, p.43.

(28) _Natural Gas_, House of Lords Select Committee of the European Communities, Session 1981-82. London: HMSO, 29 June 1982, p.18. (Henceforth cited as House of Lords).

(29) UK Department of Energy Press Notice, 10 February 1982.

(30) House of Lords, op.cit., p.53.

(31) 'Algeria's LNG Deal with UK Avoids Price Index Issue', _Petroleum Intelligence Weekly_, 5 January 1981. Neil Sinclair, 'British Gas May Rejoin Spot Market for LNG', _Lloyds List_, 27 July 1983.

(32) House of Lords, _op.cit._, p.103.

(33) _Ibid._

(34) Health and Safety Executive, _Canvey: An Investigation of Potential Hazards from Operations in the Canvey Island/Thurrock Area._ London: HMSO, 1978. _Canvey: A Second Report._ London: HMSO, 1981. Department of the Environment, _The British Gas Methane Terminal on Canvey Island_, 1983.

(35) Stern, _International Gas Trade_, Op.cit., Chapter 5.

(36) 'British Gas Demonstrate Extended Gas Making Run at Westfield in Scotland', British Gas Press Information, 12 December 1981.

(37) Commission on Energy and the Environment, _Coal and the Environment._ London: HMSO, 1981, p.123; $50 per barrel is a Shell estimate from House of Lords, p.32.

(38) _Coal and the Environment_, op.cit, p.130.

(39) _Ibid._, p.131.

(40) _House of Lords_, op.cit, p.104.

(41) 'The £20 Billion Gasbags', _The Economist_, 16 April 1983, p.47.

(42) The issue of price in West European gas trades is discussed in some detail in Stern, _International Gas Trade_, op.cit., Chapter 6.

(43) Lovegrove, _op-cit._, p.30.

(44) Ray Dafter, 'British Gas Beats off ICI', _Financial Times_, 19 September 1983.

(45) Norwegian Royal Ministry of Petroleum and Energy, Landing of Gas from the Stafjord Field, Storting Report No. 102, (1980-81), Oslo, 1981, p.50.

(46) It is possible to make this rough calculation using Department of Energy, Digest of United Kingdom Energy Statistics, 1983, Table 73, p.101. For prices in early Southern Basin contracts see Williams op.cit., pp.208-10 and Lovegrove op.cit., pp.30-33.

(47) Digest of UK Energy Statisitics, op.cit.

(48) For an explanation of this concept and an account of Norwegian natural gas export policy, see Stern, International Gas Trade, op.cit., Chapter 2.

(49) Henrik Ager-Hanssen, 'The West European Gas Market', Statoil Background Paper No 13. March 1983, p.16.

(50) 'Algeria Wins Chief Goals in Gas Deal with France', Petroleum Intelligence Weekly, 8 February 1982. This was agreed at the beginning of 1982 and since the price escalates with a basket of crude oils and fuel oils, the mid-1983 contract price would be somewhat lower. There is also a minimum price provision, and the price is delineated in French francs, but for illustrative purposes we shall ignore such complexities here.

(51) This assumes that the understanding will be adhered to (which is by no means certain) and that the gas consumption of these countries will not increase quickly in the coming decade.

(52) 'Algeria's LNG Deal with UK Avoids Price Issue', Petroleum Intelligence Weekly, ·5 January 1981.

(53) House of Lords, op.cit., p.12

(54) Algerian pricing policy and the history of the Continental contract are outlined in Stern, International Gas Trade, op.cit., Chapter 4.

(55) Hinde, op.cit., p.109.

(56) BGC, Annual Report and Accounts, 1982/83, pp.10-13.

(57) Deloitte, Haskins and Sells, British Gas Efficiency Study for the British Gas Corporation and the Department of Energy, June 1983, p.50.

(58) Ibid., p.36.

(59) Ibid., p.53.

(60) Ibid., p.50.

(61) This is a conclusion also reached in Energy Research Group, Open University, A Report on the Gas

Supply Industry, Research Report ERG0 27, March 1979, p.158.

(62) Department of Energy, Proof of Evidence for the Sizewell 'B' Public Inquiry. London: Department of Energy, 1982.

(63) Energy Policy: a Consultative Document, op.cit., p.42.

(64) Proof of Evidence, op.cit., p.A34.

(65) Ibid., p.A38.

(66) House of Lords, op.cit., p.18.

(67) BGC Annual Report and Accounts 1982/83, p.17; the quotation from the 1978 Green Paper (p.17) for HMG's reaction.

(68) This appears to be roughly the basis for the Department of Energy scenario where all non-energy (ie petrochemical) uses of gas are phased out by the first decade of the next century and the price gap between gas and coal in the industrial sector opens considerably.

(69) Burckhard Bergmann, The European Market for Natural Gas: Transition to New Orbits of Supply and Demand?, European Gas Conference '83, Oslo, 7 June 1983.

(70) United Nations Economic Commission for Europe, An Efficient Energy Future: Prospects for Europe and North America. London: Butterworth, 1983, p.38.

(71) 3rd Report from the Select Committee on Energy, Session 1980-81, The Gas Industry in Northern Ireland. London: HMSO, 30 July 1981. The information in this paragraph is drawn from that report.

(72) Ibid., Appendix 6, pp.28-30.

(73) Ibid., p.19.

(74) M. McGurnaghan and S. Scott, Trade and Cooperation in Electricity and Gas, Understanding and Cooperation in Ireland, Cooperation North, Paper 4. Belfast and Dublin: Cooperation North, July 1981.

(75) Brendan Keenan and Maurice Samuelson, 'Dublin and Belfast to be Linked by Gas Pipeline', Financial Times, 11 October 1983.

(76) Ibid., Also Maurice Samuelson, 'King Coal and the Ulster Connection', Financial Times, 30 November 1983.

(77) Brendan Keenan, 'Irish Plan Offshore Gas Pipes to Foil Terrorists', Financial Times, 13 April 1982.

(78) UK storage capacity has expanded rapidly with eight peak shaving plants and the salt cavity facility at Hornsea. See A Report on the Gas

Supply Industry, op.cit., pp.59-61. For instances
of Norwegian supply disruption, see Jonathan P.
Stern, 'Gas for Western Europe: The Choices for the
1990s', The World Today, July/August 1982,
pp. 305-14.

Table 1   Estimates of UKCS Natural Gas Reserves (BCM)

Organisation	Date of estimate	Description of reserve figure	Reserves
1. UK Department of Energy	31 December 1982	Remaining proven	633
		Remaining probable	303
		Total remaining recoverable	1,025–1,700
2. British Gas Corporation	May 1982	Remaining proven	736
		Total remaining economically and technically recoverable	1,132–1,698
3. Shell	May 1982	Known and prospective discoveries	1,698–1,840
		Anticipated discoveries as yet unknown	566
4. Phillips Petroleum	May 1982	Proven and probable	1,557
		Possible	1,557–1,840
5. British Petroleum	March 1983	Contracted firm gas remaining	764
		Discovered and identified prospects	991
		Expected discoveries	425

Table 1 (cont'd)

Organisation	Date of estimate	Description of reserve figure	Reserves
6. Peter Odell	March 1982	Remaining proven/probable reserves 1981	1,075
		Likely remaining reserves discoverable by 1985	2,200

Sources:

1. Department of Energy, Development of the Oil and Gas Resources of the United Kingdom (Brown Book), 1983 Tables 2a and 2b, pp.5-6.
2. Natural Gas, Report of House of Lords Select Committee of the European Communities. London: HMSO, 20 June 1982, p.106.
3. Ibid., p.35.
4. Ibid., p.68.
5. J. B. B. Bullough, UK Gas Future – Operators Viewpoint, HYDROCARBONS '83 Conference, Great Yarmouth, 1-3 March 1983.
6. Energy Advice, Alternative Strategies for Natural Gas in Western Europe. Geneva: Energy Advice, 1982, Table 1, p.39

Table 2   Estimates of Recoverable Gas Reserves in Present Discoveries on the UKCS as at 31 December 1982 (BCM)

	Proven	Probable	Proven+ Probable	Possible
Initial recoverable reserves:				
From dry gas fields:				
1. In production or under development				
a. Southern Basin	646	31	677	14
b. UK Frigg and Morecambe	198	28	226	34
	844	59	903	48
2. Other significant discoveries not yet fully appraised				
a. Southern Basin	48	42	90	51
b. Other areas	–	–	–	42
Total dry gas	892	101	993	141
From gas condensate fields:(a)				
1. In production or under development	23	6	29	–

Table 2 (cont'd 1)

	Proven	Probable	Proven+ Probable	Possible
2. Other significant discoveries not yet fully appraised	17	153	170	337
	40̄	1̄5̄9̄	1̄9̄9̄(b)	3̄3̄7̄
Associated gas from oil fields:(a)				
1. In production or under development:				
a. Currently delivering gas ashore	91	3	94	3
b. Expected to be connected	40	14	54	8
Sub-total	1̄3̄1̄	1̄7̄	1̄4̄8̄	1̄1̄
2. Other significant discoveries not yet fully appraised	3	31	34	37
Total associated gas	1̄3̄4̄	4̄8̄	1̄8̄2̄	4̄8̄
Total initial reserves in present discoveries	1,066	308	1,374	526

67

Table 2 (cont'd 2)

	Proven	Probable	Proven+ Probable	Possible
Remaining recoverable reserves:				
Cumulative production to end of 1982 (c)				
1. Dry gas				
a. Southern Basin	395			
b. UK Frigg and Morecambe	29			
2. Associated gas from oil fields	9			
Total cumulative production to end of 1982	433			
Total remaining reserves in present discoveries	633	308	941	

(a) All in Northern Sector of North Sea.
(b) Figure of 994 in source is an error.
(c) Excludes flared gas.

Source: Ibid., Table 2a, p.5.

Table 3   Range of Initial Recoverable Gas Reserves on the UKCS(a) (BCM)

1. Reserves in present discoveries

a.	dry gas fields	957-1,053
b.	gas condensate fields	125 – 314
c.	gas associated with oil	164 – 198
	Total	1,246-1,565

2. Reserves in potential future discoveries

a.	gas prospects	60 – 575
b.	gas condensate prospects	unassessable
c.	gas associated with oil	5 – 60
	Total	65 – 635

3. Statistically aggregated range of initial recoverable reserves on UKCS (b)   1,450-2,125

4. Remaining recoverable reserves on UKCS (c)   1,017-1,692

(a) No allowance has been made for possible accumulations of gas at deeper levels in existing discoveries or areas which have not yet been drilled.
(b) Statistical aggregation which results in a narrowing of range of reserves, cf. 1982 estimate of 1,350-2,250 BCM.
(c) Arrived at by subtracting 433 BCM of gas already produced from figures in 3.

Source: Ibid., Table 2b, p.6.

Table 4   Gas Production from UKCS (million cubic metres) 1976-82

Southern Basin Fields	Total to end 1976	1977	1978	1979	1980	1981	1982	Total to end 1982
West Sole	16,207	1,947	1,533	1,365	1,445	1,455	1,512	25,464
Leman Bank	96,730	15,581	14,719	13,831	9,482	13,207	11,675	175,225
Hewett Area	39,565	7,852	6,392	6,288	6,568	5,048	4,108	75,821
Indefatigable	27,377	6,779	6,450	6,006	6,878	5,613	5,720	64,823
Viking Area	21,311	6,330	5,238	4,397	4,689	3,307	4,381	49,653
Rough	522	1,063	931	1,005	467	99	101	4,188
Frigg(a)		614	2,907	5,345	6,374	7,057	6,569	28,866
Piper(b)			4	536	521	520	629	2,210
FLAGS(c)							2,144	2,144
Other(d)	10	138	323	455	866	1,098	1,437	4,327
Total(e)	201,722	40,304	38,497	39,228	37,290	37,404	38,276	432,721

Table 4 (cont'd)

(a) UK share only
(b) Gas used offshore or delivered to land via the
    Frigg pipeline system.
(c) Gas delivered to land via the Far North Liquid
    and Associated Gas (FLAG) System.
(d) Associated gas, mainly methane, produced and
    used mainly on Northern Basin oil poduction
    platforms.
(e) Gross production, ie includes own use for
    drilling, production and pumping, but excludes
    gas flared.

Sources: Ibid., Appendix 9, p.53.

Table 5 Gas Flaring at Oilfields and Terminals (million cubic metres) 1978-82

	1978	1979	1980	1981	1982	Total to end 1982
Argyll	40	53	49	30	75	377
Auk	47	35	23	27	27	281
Beatrice	–	–	–	9	31	40
Beryl	247	201	138	102	108	1,683
Brent	1,598	3,347	1,264	1,221	1,123	9,120
Buchan	–	–	–	59	86	145
Claymore	120	41	12	14	14	208
South Cormorant	–	5	151	80	56	292
North Cormorant	–	–	–	–	36	36
Dunlin	34	265	224	181	95	799
Forties	1,172	1,129	911	798	738	6,321
Fulmar	–	–	–	–	289	289
Heather	19	99	35	83	109	345

Table 5 (cont'd)	1978	1979	1980	1981	1982	Total to end 1982
Montrose	182	200	183	151	117	979
Murchison(a)	–	–	47	381	217	582
Ninian	3	453	552	485	213	1,706
Piper	933	578	310	185	135	2,906
Statfjord(b)	–	9	72	19	32	132
Tartan	–	–	–	124	86	210
Thistle	151	189	223	209	218	990
Total offshore	4,546	6,604	4,194	4,095	3,805	27,441
Wytch Farm	–	2	7	2	1	12
Flotta Terminal(c)	–	–	44	34	17	95
Sullom Voe Terminal(d)	–	–	–	87	226	313
Total onshore	–	2	51	123	244	420
Total gas flared(e)	4,546	6,606	4,245	4,218	4,049	27,861

Notes to Table 5

(a)   UK share (83.75%) of wellhead production.
(b)   UK share (15.9069%) of wellhead production.
(c)   Pipeline terminal serving Claymore, Piper and
Tartan fields.
(d)   Pipeline terminal serving Brent, North and
South Cormorant, Dunlin, Heather, Murchison, Ninian
and Thistle fields.
(e)   Gaseous hydrocarbons associated with crude oil
production containing methane, ethane, propane,
butane and condensates.

Sources: UK Department of Energy, Digest of United
Kingdom Energy Statistics, 1983.   London: HMSO,
1983, Table 30, p.45.

Table 6   Cost Structure of BGC 1974-84

	1974-75	1975-76	1976-77	1977-78	1978-79	1978-80	1980-81	1981-82	1982-83	1983-84*
Natural gas and feedstock pence per therm	1.68	1.90	2.06	3.07	4.65	5.78	7.87	10.62	11.64	13.61
% of total cost	20.0	18.5	16.2	20.7	29.3	31.2	33.0	36.4	36.2	40.2
Operating costs %	80.0	81.5	83.8	79.3	70.7	68.8	63.8	55.6	53.9	50.8
Gas levy %							3.3	8.0	9.9	9.1

* estimate

Source: British Gas Corporation, Annual Report and Accounts, 1982-83, pp.22-3, 28-9.

75

Table 7

The Cost of Bringing New Norwegian Gas to the Market

	(1983) $ per mmbtu	pence per therm
Field	4.00-6.00	25-40
Transportation	1.50-2.50	9-16
	5.50-8.50	34-56

Source: Henrik Ager-Hanssen, 'The West European
Gas Market', Statoil Background paper No.13. March
1983, Fig.19.

Table 8

Cost/Price Reference Frame for Domestic and Imported Base Load Sources of Supply

	$ per mmbtu	pence per therm
Average cost of BGC natural gas supplies 1982-83	1.86	11.64
Estimated average price paid for UKCS gas in 1982	1.37	8.57
Estimated price paid for Frigg gas in 1982	3.60	22.50
1980 contract price for Statfjord gas landed in Continental Europe	5.50	34.4
Norwegian estimate of field development and transportation cost for new Norwegian gas (not including government take) 1983	5.50 - 8.50	34 - 56
Siberian gas landed in UK (estimated French 1982 contract price plus allowance for additional transportation to UK)	5.15 - 5.25	32 - 33
Algerian gas landed at new UK LNG terminal (as yet unbuilt), based on 1982 French contract with $35 per bbl oil price	5.90	36.9
SNG from coal (1982 estimate)	8.62	53.9

Source: 1982-83 gas supply cost from Table 6, Norwegian estimate from Table 7, all other figures are illustrative and estimated by the author from fragmentary published material.

Table 9   UK Natural Gas Demand, Scenario 'BL'(a)

	1990	2000	2010
	%	%	%
Domestic	56.4	65.7	71.7
Iron and steel	1.7	1.2	0.6
Other industry(b)	28.7	18.3	12.6
Other consumers	12.7	14.8	15.7
Total (BCM)	49.8	46.5	43.7

(a)   The 'BL' scenario is one of eight developed by
the Department of Energy.   This scenario is based on
the assumptions of low oil/energy pricess, low
industrial growth and moderate (1½ per cent per
annum) growth of GDP.

(b)   This includes non-energy uses of natural gas,
mainly petrochemicals.

    The percentages do not add exactly to 100
because of rounding and conversion error.

Source: UK Department of Energy, Proof of Evidence
for the Sizewell 'B' Public Inquiry, October 1982,
Tables D and F, pp.44-6.

Table 10   UK Natural Gas Price Scenario 'BL'

1980 = 100 (figures in brackets are prices in pence per therm)

	1990	2000	2010
Domestic			
Natural gas	165 (37.1)	211 (47.5)	275 (61.9)
Oil	99 (38.6)	158 (61.6)	203 (79.2)
Coal	111 (26.1)	150 (35.3)	176 (41.4)
Electricity (average consumer 750–7500 Kwh pa)	100	129	143
Industrial			
Natural gas	118 (20.8)	190 (33.4)	289 (50.9)
Fuel oil	100 (22.5)	169 (38.0)	220 (49.5)
Gas oil	99 (34.7)	160 (56.0)	206 (72.1)
Coal	107 (15.6)	160 (23.4)	190 (27.7)
Electricity	109 (75.6)	156 (108.2)	176 (122.1)

Table 10 (cont'd)

	1990	2000	2010
Private services			
Natural gas	136 (33.0)	179 (43.3)	238 (57.8)
Oil	99 (35.6)	158 (56.9)	203 (73.1)
Solid fuels	113 (41.4)	144 (52.7)	169 (61.9)
Electricity	104	137	153
Public services			
Natural gas	136 (33.0)	179 (43.5)	238 (57.8)
Oil	99 (35.6)	158 (56.9)	203 (49.5)
Coal	106 (17.3)	149 (24.3)	174 (28.4)
Electricity	104	139	155
Crude oil price $ bbl	27	43	55

Source: Department of Energy, Proof of Evidence for the Sizewell 'B' Public Inquiry, October 1982, Table A, P.A40.

Table 11   Sectoral Consumption of Natural Gas* (%)

	Power stations	Iron and steel	Other industry	Domestic	Public admin. and commercial
1977	3.6	3.3	37.0	45.2	10.9
1978	2.2	2.9	36.1	47.4	11.4
1979	1.4	3.2	33.9	49.6	11.9
1980	0.8	2.7	33.3	50.7	12.4
1981	0.5	2.5	31.7	52.7	12.7
1982	0.5	2.2	31.9	52.3	13.1
1983**	0.5	2.0	31.9	52.4	13.3

* In 1977 town gas amounted to 0.5% of UK gas supply; by 1982 this had fallen to 0.2%.
** Three quarters only

Source: Department of Energy, Energy Trends, June 1983, Table 8.

Table 12   Total UK Primary Fuel Consumption and the Natural Gas Position (a)

	1978	1979	1980	1981	1982
Coal	33.3	34.4	35.1	35.5	33.9
Petroleum	44.4	42.5	39.9	38.1	38.9
Natural gas	18.1	18.9	20.6	21.6	21.6
Nuclear power	3.7	3.7	3.8	4.1	4.9
Hydro power	0.6	0.6	0.6	0.7	0.7
Total (mtoe)	211.9	221.4	202.7	196.2	192.6
Natural gas(b) consumption (mtoe)	38.3	41.8	41.8	42.4	41.6
Natural gas production as a % of consumption	88.5	81.8	76.6	75.7	77.9

(a) includes oil and gas for non-energy use and marine bunkers.
(b) includes land and colliery methane and associated gas produced and used mainly on Northern sector oil production platforms. Excludes gas flared or reinjected.

Source: Brown Book 1983 adapted from Table 8, p.22

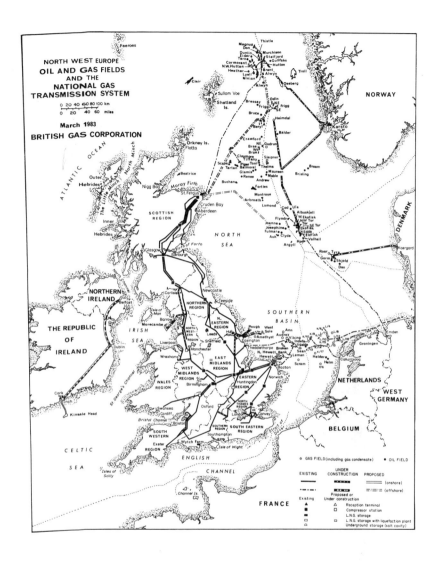

NORTH WEST EUROPE
OIL AND GAS FIELDS
AND THE
**NATIONAL GAS
TRANSMISSION SYSTEM**

0 20 40 60 80 100 km
0   20   40   60 miles

March 1983
BRITISH GAS CORPORATION

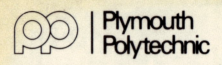

Plymouth
Polytechnic

Learning Resources Centre